THE HISTORY OF DORMAN SMITH

1878–1972

Front endpaper: MACHINE SHOP circa 1900

Rear endpaper: ASSEMBLY SHOP 1972

The History of Dorman Smith

1878–1972

By

NORMAN LEE

Senior Lecturer in Economics in the University of Manchester

and

PETER STUBBS

Senior Lecturer in Economics in the University of Manchester

NEWMAN NEAME

ERRATA

Page 68 line 10. For *Gambridge* read *Sambidge.*
Page 116 line 9. For *sales manager* read *sales director.*
Page 119 line 3. For *contact breaker* read *circuit breaker.*
Page 157 line 8. For *Hoffman* read *Hoffmann.*

SBN 08 017414 0

And he gave it for his opinion, that whoever
could make two ears of corn or two blades
of grass to grow upon a spot of ground
where only one grew before, would deserve better
of mankind, and do more essential service
to his country, than the whole race of
politicians put together.

Jonathan Swift

First published in 1972 by Newman Neame Limited, London
Printed in Great Britain by A. Wheaton & Co, Exeter

This book is set in 12 on 13 point Bembo

Table of Contents

Illustrations

List of Tables

Authors' Preface

We undertook this study because it concerned a company which has a long and interesting history in the electrical engineering industry although it remains quite modest in size by contemporary standards. It is now one of the oldest surviving members of the industry, since practically all the companies established in Britain at the same time have failed or been changed out of all recognition by mergers. Moreover, the company today is as vigorous as it ever was. The study offered a unique opportunity to trace the vicissitudes of a company across almost a century, in which time both company and industry grew from infancy to maturity. The company is not large by the standards of the industry, but it is nonetheless interesting for there are many histories of large companies but comparatively few of medium size enterprises; the latter have been relatively neglected in business histories.

As we pursued our investigation into the company's past we were not disappointed, for its growth has not been smooth and easy. The influence of personalities has been strong and has caused significant ebbs and flows in the rate of growth of the firm, quite apart from all the external factors, such as world wars, which have affected all members of the industry. This influence not only involves the changes of command in the company, but also the mutation of individual personalities over time, as Charles Dorman and Reginald Smith retreated from the ambitious expansion of their youth into financial conservatism as they aged, and Herbert Baggs spent his latter years as chairman preoccupied with engineering problems while neglecting the commercial side of the business. For the authors, whose professional interests are concerned with reducing social behaviour into measurable economic variables, the work has shown how critically personalities can influence both the short and long-run fortunes of the individual business. Yet the theory of the firm in economics takes relatively little account of such influences.

Dr Stubbs was basically responsible for Chapters One, Four and Five, and Dr Lee for Chapters Two and Three. In Chapter One, the background to the industry is sketched, since it has not received much attention in economic literature, and the life of the founder, John Raworth – a Yorkshireman who started this eminently Lancastrian company. Raworth is interesting as an innovator, a man raised in the tradition of steam technology who was able to switch with great success to the new medium of electricity. Chapter Two records how his two young protégés, Charles Dorman and Reginald Smith, bought Raworth's business after the government of the day had legislated the industry into virtual stagnation. They were quickly successful, actively introducing new products so that they gave up installation work and became specialist manufacturers with a new factory. In latter years their energies seemed to flag and their successor, Herbert Baggs, was much more an engineer than a business man, as seen in Chapter Three. Consequently, while the industry advanced enormously between the wars the company marked time until its secretary, Thomas Atherton, gambled everything on realising the full potential of the company, borrowed heavily from a bank and bought out the interest of Baggs, who was anxious to retire. Atherton's success is examined in Chapter Four – the restoration of confidence among the executives, the expansion of capacity during the war and the development of the fused plug for the post-war market. The development of the mature holding company follows in Chapter Five: opposition by rivals brought disappointment over the fused plug and profits in some subsidiaries flagged, but on the credit side the company's range of products became steadily wider and more sophisticated, consultants helped to raise managerial standards and the move to a much bigger modern factory helped to transform the group. The potentially dangerous issue of managerial succession as Thomas Atherton approached retirement was solved by the growing success of his son, Geoffrey, under whose direction the group has since achieved record levels of profit and profitability.

We owe thanks to several people for invaluable help. Above all we have received unstinting co-operation from the company itself: executives past and present have given valuable time and thought to our questions, which they treated with great candour. In particular we should like to thank Thomas Atherton and his sons, Geoffrey and Eric; Robert Dale, Myles L. Cooper, Keith Blackshaw and George Wood, as well as a large number of individuals in the company who settled various points of accuracy. Outside the company, the Institution of Electrical Engineers generously opened its library to us, and Neil Bright and Simon Pine helped to search the literature in Manchester and collate some of the accounting data. Mr G. R. C. McDowell, of George H. Scholes & Company Limited, lent us material on the early days of the industry, and Professor Dennis Coppock and Mr Roy Wallis of the University of Manchester commented usefully on parts of the draft manuscript. Misses Kathleen Deignan, Margot Cornock (now Mrs Margot Tyson) and Alwyn Newton typed the manuscripts and Mrs Jennifer Buchanan the final draft.

CHAPTER ONE

John Raworth, Founder

His birth and boyhood

John Smith Raworth* (*Plate 1*) was born in Sheffield on 16 February 1846,[1] to B. J. Raworth and his wife, Epenetes. The family had been well known for several generations in the iron and steel trade, and John showed an early interest in mechanics. At Sheffield he went to grammar school but his father died before John and his brothers had finished school. They went to live at Tapton House in Chesterfield in 1858. His two younger brothers, Alfred, aged nine, and Henry, aged eleven, entered Chesterfield School in that year, but John did not enter until 1861 when he was just fifteen.[2] It is interesting that Tapton House had once been the home of George Stephenson, the railway pioneer who built the famous *Rocket* locomotive. It would take a very dull boy not to be inspired by such an association: and John and Alfred were by no means dull boys, both eventually rising to the highest ranks of their engineering professions. About their home life we can only speculate, but their industry in later life and John's Biblical allusions in his letters and speeches make it clear that the influence of their Wesleyan Methodist mother was strong. However, at sixteen John left home to be apprenticed for three years to Mr Edward Hayes of Stony Stratford in Buckinghamshire.

* Occasional spelling errors by reporters make it clear that his name was pronounced Ray-worth.

[1] The published obituaries of Raworth give conflicting dates of birth: *The Engineer* cited 1 June 1846, but *Engineering*, the journal edited by Raworth's brother, Benjamin Alfred, cited 16 February. The latter is presumably correct, and *The Engineer* could well have mistakenly consulted Benjamin's biography, as his birthday was 1 June.

[2] The actual entry dates to Chesterfield School were: Benjamin Alfred, 1 June 1858; Henry, 22 October 1858; John, 16 February 1861. We are indebted to Joan Goodwin, Secretary at Chesterfield School, for this information.

His brother Alfred had a more academic start, going on from Chesterfield School to Owens College in Manchester, although later the two brothers were to come together again under a common employer.

The origins of the electrical industry

At this stage it may help our perspective to look at the development of the electrical industry and its place within industry as a whole. By the time of Raworth's apprenticeship, the most obvious and far-reaching transformation of the nineteenth century had been wrought by the steam engine; there were about ten thousand miles of railways in Britain, and they were spreading rapidly across the United States, spanning the continent by 1869; at sea, steam propulsion was also creating a revolution. At this time, the electricity industry was very much in its infancy, and few could foresee that it would have just as much influence, and eventually more, on human life.

Scientific investigation of electrical phenomena[3] went back to antiquity, with the Greek discovery of static electricity when amber was rubbed with a dry cloth. Later, Englishmen made their contribution. Roger Bacon, in his *Opus Tertium* of 1267, described the identification and properties of the lodestone and the determination of its poles. The Court Physician to Queen Elizabeth the First, Dr William Gilbert, coined the word 'electricity', publishing his *De Magnete* in 1600.

In the eighteenth century the Leyden jar was invented, which enabled static electricity to be accumulated. The celebrated American, Benjamin Franklin, experimented with Leyden jars to study sparks and, after flying kites in thunderstorms, in 1752 he developed the first practical electrical invention, the lightning conductor. Leyden jars were also used in 1747 by a group of English scientists in London to demonstrate the transmission of electricity along a two and a half mile wire. However, the great

[3] There are many histories available of electrical discovery. A comprehensive account is given by Dr Percy Dunsheath, *A History of Electrical Engineering*, Faber, London, 1962.

PLATE 1. John Raworth and his employees about 1884. Raworth is in the middle of the second row, with C. M. Dorman on his extreme right and R. A. Smith seated between them. Fourth from left in the back row is R. A. Dawbarn, with G. Wilkinson on his right

PLATE 2. Raworth's early premises at 22–24 Brazennose Street, Manchester

stumbling block for all these pioneers was their inability to produce a continuous flow of electrical current; not until the battery was invented could there be any real progress.

It is one of the curiosities of electrical science that frogs' legs should figure in its history. In Bologna, Professor Luigi Galvani found that by connecting the legs of a newly killed frog with a metallic conductor, they could be made to twitch, and twitch still more if two different metals were used, one touching the nerve, the other the muscle. This discovery excited the scientific world, and gave rise to the expression *galvanism*. Professor Alessandro Volta at Pavia repeated Galvani's experiments and went on to make one of the classic discoveries. In March 1800 he sent a paper to the Royal Society in London entitled 'On the Electricity Excited by the Mere Contact of Conducting Substances of Different Kinds'; the battery was invented, and though its full significance was not appreciated, experimenters at last had a steady source of electrical power from copper-zinc-acid cells. One of the most outstanding pioneers was the young Humphry Davy, who within a few years of Volta's discovery had discovered the electric arc and, using electrolysis, had discovered and named potassium, sodium, barium, strontium, calcium and magnesium. However, the early batteries were crude, and only after local action had been countered by amalgamating the zinc plate with mercury, and Daniell invented the self-depolarising cell, were long lasting and reliable batteries available. In 1868 the Leclanché cell was invented, and the contemporary dry cell battery still embodies its principles.

However, important though the battery was, it was not destined to become the most important source of electric power. In the early nineteenth century a number of scientists who are now famous – Oersted, Ampère, Henry and Davy's assistant, Faraday, – were gradually unveiling the secrets of electromagnetism. An electric current was found to induce magnetism in an iron bar; later the phenomenon was found to be reversible and a magnet could induce an electric current in nearby wire coils. The principles which governed electric power generation were revealed.

The four most important electrical inventions perfected in the nineteenth century were the electric telegraph, the electric light, the generation of electric power and electric traction. All these developments were to impinge in one way or another on John Raworth, and in several of them he was to make significant contributions.

The coming of the electric telegraph

The telegraph was the first electrical device to provide the basis for an industry. It had many pioneers but the crucial breakthrough came on 28 July 1837, when William Cooke and Professor Wheatstone transmitted messages between Euston and Camden Town stations on the London and North Western Railway. Two years later the Great Western Railway Company installed the first commercial telegraph along the thirteen miles between Paddington and West Drayton, soon extending it to Slough. In 1842 the telegraph suddenly caught the public imagination when it provided the crucial link in a murder case: the suspect boarded the train at Slough; his description was wired to Paddington, where he was followed on arrival and arrested. Thereafter the growth of the telegraph system was rapid. The first commercial submarine cable was laid between Dover and Calais in 1850, only to be severed by a fisherman from Boulogne,[4] but it was relaid in the following year with a much stronger cable, and the service was opened to the public on 19 October 1851. The traffic was so heavy that within two years a second cable had to be laid. While Raworth was still at school, New York and San Francisco were linked by telegraph and by the time he finished his second apprenticeship in 1868 there were over sixteen thousand miles of telegraph in Britain.

At this stage there were very few firms engaged in an electrical industry as such. All that a telegraph system required was a cable and a few simple instruments, and batteries to provide power. In contrast, the world of steam power involved an immense industrial complex, making locomotives, rolling stock, rails, steamships and a host of auxiliary machines. In the mid-1860s

[4] The fisherman was apparently the first electrical thief. On raising the cable, he mistook the copper core for gold and made off with a considerable length of it, according to Charles Bright, *Submarine Telegraphs*, London, 1898.

when Raworth completed his apprenticeship at Stony Stratford the British railway system was still expanding, and there was a very big market for its products both at home and abroad. An economic historian of the period observed:[5]

> British engineering in the 1860s was still dominated by those sectors which had emerged from the industrial revolution earlier in the century – textile machinery, railway rolling stock, steam engines and boilers of all kinds and the relevant, mostly heavy, machine tools. During the third and fourth decades of the century the industry had begun to develop away from the earlier pattern, whereby all manner of metal jobbing was undertaken, into more specialised organisations. Complete specialisation was still rare but makers of locomotives, textile machinery and heavy machine tools came closest to it . . . The older locomotive firms such as Stephenson, Hawthorn and Sharp Stewart still did considerable business in machine tools, steam engines for all uses and civil engineering.

Raworth's formative years

From Stony Stratford, Raworth moved to the firm just mentioned, R. and W. Hawthorn of Newcastle upon Tyne. It was a big firm for its time, employing over a thousand men,[6] and Raworth spent two years there working in the shops and the drawing office, gaining experience which was to stand him in excellent stead as his career progressed.

In 1868 Raworth joined the Manchester engineering firm of Wren and Hopkinson, who described themselves as 'millwrights, engineers, machine makers, iron and brass founders, manufac-

[5] S. B. Saul, 'The Engineering Industry', in D. H. Aldcroft, *The Development of British Industry and Foreign Competition, 1875–1914*, Allen and Unwin, London, 1968, p. 186.

[6] S. B. Saul, 'The Market and the Development of the Mechanical Engineering Industries in Britain, 1860–1914', *Economic History Review*, Vol. 20, April 1967, p. 111–130.

turers of steam engines, hydraulic presses and hoisting machinery'. The Hopkinsons were a notable Manchester family, while the Wren side of the business traced its lineage back to Boulton and Watt, with whom they had associated as Manchester District agents for the early Watt steam engines. Raworth himself was a keen champion of the steam engine, and later was to design a novel engine himself and observe:[7]

> The main muscle of civilisation is the steam engine. Steam supplies the power and man directs it. The price at which steam power can be supplied is the principal factor in determining at any moment the degree of domination which man may exercise over the world in which he lives.

Raworth joined Wren and Hopkinson as a draughtsman but he soon showed that he had greater talents, and Alderman John Hopkinson chose him as his assistant. The Hopkinsons were a family of great ability, and there can be no doubt that Raworth was deeply influenced by his experience as their employee; perhaps he played a part in persuading his brother Alfred to work for Wren and Hopkinson after Alfred graduated from Owens College where he had been a contemporary of Hopkinson's brilliant son, John. An authoritative source notes:[8]

> John Hopkinson, with his sons, had a very large influence on the development, from the practical as well as the theoretical standpoint, of electrical machinery. Mr Raworth was thus brought into intimate association with the beginning of great things, and his training as a mechanical engineer enabled him to approach problems from a broader standpoint than was then the case with many electrical engineers.

[7] J. S. Raworth, 'Cheap Steam Power', *Proceedings of the Northern Society of Electrical Engineers*, Vol. 3, No. 2, 1896, pp. 1–23.

[8] This account was given in John Raworth's obituary in *Engineering*, 30 March 1917, p. 398. As the journal was edited by Raworth's brother, Alfred, its authenticity is strong.

The Hopkinsons were a remarkable family, and one of the sons, Sir Alfred Hopkinson, has left behind an interesting account of his experiences,[9] in which he says:

> The engineering firm of which my father was the active member kept fully abreast of the practical engineering knowledge of the time and none could have been more ready to adopt and apply new notions.

Sir Alfred Hopkinson was a King's Counsel, Member of Parliament, and Vice-Chancellor of Manchester University, and three other sons of Alderman John Hopkinson became successful engineers. Another son, John Hopkinson, was a mathematician; he took a three-year course in electrical engineering at Owens College, Manchester, entering in 1865 at the age of sixteen and going on to Trinity College, Cambridge, in 1867 to become Senior Wrangler of the Mathematics Tripos in 1871. Not only was he academically eminent, but he also pioneered the admission of non-conformists into Cambridge:

> Hopkinson came of that hard-working, middle-class, non-conformist stock, with its deep religious convictions and high intellectual ambitions, which was such a striking social feature of industrial England during the latter part of the nineteenth century. His father was a Lancashire millwright who by dint of hard work and ability had risen to be partner in a long-established and important Manchester firm of engineers . . . Hopkinson's mother must have been an equally remarkable person for she produced a family of thirteen children, several of whom achieved outstanding careers: John, the eldest, became the electrical engineer, while Alfred, the second son, rose to a position of eminence as a KC, Member of Parliament and Vice-Chancellor of a University. Both

[9] Sir Alfred Hopkinson, *Penultima*, Martin Hopkinson, London, 1930.

> father and mother played a vital part in the intellectual achievements of their brilliant children. The father, always interested in the developments of electricity going on at the time, set up simple experiments in the attic and so secured John's early interest.[10]

After the young John Hopkinson graduated at Cambridge in 1871, he was the first non-conformist to be elected to a Fellowship. However, he went down and for a few months worked at the Wren and Hopkinson factory while Raworth was employed there. We cannot tell whether the younger John had much influence on Raworth but the father, Alderman John, undoubtedly did. During his four years with Hopkinson, Raworth did much original work

> but more important still, he laid the foundation by the further acquisition of experience, for the work of greater originality in later years.[11]

Manchester at that time was prosperous and self-confident.[12] National production and exports were buoyant and the industrial north was the powerhouse of England. The Hopkinson family and literally hundreds of thousands of others read Samuel Smiles's *Self-Help* and *Lives of the Engineers*, and absorbed their lessons. In 1872 John Raworth, whose confidence in his own ability must have been rising steadily, decided to strike out on his own account as a designer and manufacturer of machinery for cotton spinning and weaving.

There are no records left of this period of Raworth's career, and we can only surmise about the progress of his business. Patent records show J. S. Raworth applying in 1875 for patents for weaving handkerchiefs, and doubling and winding cotton; in 1874 for spanning and doubling, and in 1875 for beetling fabrics. In 1875 there is also a joint application with his brother, Alfred, for winding yarns and another in 1877 for looms. After leaving

[10] Dunsheath, *op. cit.*, pp. 114–115.
[11] *Engineering*, loc. cit.
[12] See Asa Briggs, *Victorian Cities*, Penguin Books, 1968, Chapter 3.

Wren and Hopkinson, Alfred had been departmental works manager with Siemens at Woolwich, and he then was said to be 'in practice as an engineer in Manchester'. We do not know how close his commercial association with John was, but they were obviously in technical association. However, the timing for John's venture into independent business could hardly have been less fortunate, for in 1873 the British economy entered a depression which was to last until 1879. Such a depression would have been particularly damaging to a manufacturer of capital equipment like Raworth, because manufacturers postponed their orders for new machines. In a trade directory of 1877–8,[13] we find Raworth listed as an 'engineer and machinist' at 13 Saint Mary's Gate, Manchester, and in the 1879[14] directory there is listed 'J. S. Raworth and Company, doublers and manufacturers of sponge cloth, towels etc' at Wellington Mills, Piercy Street, Ancoats, Manchester. His fortunes can hardly have been good, for by 1881 there was no trace of either his machinery or his textile ventures, and he never referred to them in later life. However, in 1878 an opportunity arose which was to cancel out these misfortunes: he became the Lancashire and Yorkshire agent for the successful electrical enterprise, Siemens Brothers.

The contribution of the Siemens

The 1870s had seen significant advances in the practical application of electricity, with the introduction of the electric light, and the perfection of the dynamo to generate electric current. The Siemens brothers were a rare combination of inventive ability and commercial acumen, and made important and original contributions to electrical technology.[15] Werner Siemens had wanted to join the Prussian army as an engineer, but instead had been obliged to accept an officer cadetship in the artillery in 1834. After training in the Artillery and Engineers School in Berlin, he became bored with the routine of garrison life and occupied his

[13] *Isaac Slater's Directory of Manchester*, 1878.

[14] *Isaac Slater's Directory of Manchester*, 1879.

[15] The account that follows is based largely upon Georg Siemens, *History of the House of Siemens*, Vol. I, 1847–1914, Verlag Karl Alber, Munich, 1957.

spare time in scientific experiments. A friend's father had some Wheatstone telegraph instruments which he had brought across from England, and Werner Siemens saw that it would be possible to improve them. He took out a patent in 1847, and was made senior army staff member to the Prussian Telegraph Commission. He was driven not only by scientific curiosity but also by anxiety to provide for his younger brothers and sisters, since they had lost their parents. With the mechanic Halske as partner and a cousin providing capital as a sleeping partner, the firm of Siemens and Halske began business on 1 October 1847, employing ten men. Very shortly, Werner achieved an important breakthrough by sheathing telegraph cable in gutta percha, which enabled it to be laid underground. The business flourished, and Werner persuaded his brother Wilhelm in London to become Siemens and Halske's representative in England.

Throughout the 1850s and 1860s, the Siemens brothers earned admiration and profit in their cable laying enterprise. Brett and Newall had established themselves as the leading submarine telegraph layers with their first two cross Channel cables from Dover to Calais and later ones to Ireland, Holland and Germany. But cable laying in the deep waters of the Mediterranean, where depths were as much as fifteen hundred fathoms, was a much tougher proposition. Three attempts failed: if the cable was paid out too slowly, it snapped; too quickly, and the cable was spent before the destination was reached. Werner evolved a theory to cure these problems, and Siemens and Halske became Telegraph Consultants to Newall and Company. Siemens, Halske and Company was formed in London in October 1858 and set up a cable factory near Woolwich, but Halske viewed submarine telegraphs as foolish gambles, and after Siemens, Halske and Company lost £15,000, or half its capital, when the Cartagena-Oran cable snapped during laying, Halske sold out his interest in the London business, which was renamed Siemens Brothers, under Werner and Wilhelm, now naturalised and anglicised as William. The fame of the company grew with their completion of the world's longest telegraph, from London to Calcutta in 1870.

By this time, electric lighting was little further than the experimental stage. The principle of the arc light had been appreciated since Sir Humphry Davy's experiments in 1809, but its practical use seemed distant. The arc required much more current than batteries could conveniently provide, and the carbon elements vaporised and thus required permanent attendance. Cumbersome magneto-electric machines could provide the necessary current but the whole system was inordinately expensive and large and could only be justified in very special cases, notably lighthouses; electric arc lamps were first used at the South Foreland lighthouse in 1858.[16] However, in January 1867 Werner Siemens took out a patent on a machine which produced electricity without the use of permanent magnets – a self-exciting machine which the brothers called a *dynamo-electric machine*. Its current fluctuated but improvements by the Belgian, Gramme, in 1870 made the dynamo a practical source of electric power.

Siemens, Halske in Germany became leading manufacturers of dynamos in the 1870s, and the British firm of Siemens Brothers kept its lead in submarine cable technology by launching in 1874 the first ship designed specifically for cable laying, the *Faraday*. Dynamo technology improved steadily through the 1870s and 1880s. It is interesting that one of the leading contributors was Dr John Hopkinson who, as the son of Alderman Hopkinson, had worked in the same factory as Raworth in 1871. Hopkinson experimented with a Siemens dynamo and set up as an electrical consultant in 1877. Dunsheath ranks him extremely high, with Kelvin, Edison and Swan, noting:[17]

> . . . as a mathematician of a very high order, Hopkinson contributed effectively to the fundamental development of dynamo design, power transmission and alternating current working . . . This brief summary cannot convey an adequate estimate of the outstanding contribution made by Hopkinson. As both mathematician and engineer, he reached the pinnacle of achievement in his day, a fact which can only be appreciated by a study of

[16] Even so, by 1880 there were only ten electric lighthouses in the world.
[17] Dunsheath, *op. cit.*, pp. 114, 117.

> the many papers he wrote . . . The contributions to electrical engineering were unique.

Advances in electric lighting

Thanks to the work of Siemens, Gramme, Hopkinson and Edison, heavy continuous currents became available. Siemens dynamos in particular were noted for their high efficiency.[18] The dynamos made carbon arc lights much more practicable, and they came into increasing commercial use in the late 1870s. However, there were problems in using arc lights, for although it was intended that the arc should illuminate, in practice much of the light came from the incandescent tips of carbon. As the tips of carbon burned away, they had somehow to be moved together to maintain a steady gap across which the current could arc. Much Victorian ingenuity went into the solution of this problem, including clockwork devices and magnets to keep the gap tolerably constant. In 1867 arc lamps had run on trial in Paris, Philadelphia and London.

Two names are especially associated with the solution of the arc problem – Brush and Jablochkoff – though they were soon to be overshadowed by the invention of the incandescent lamp. Paul Jablochkoff was a Russian army engineer working in Paris, where he produced his famous candle in 1877.[19] It consisted of two nine-inch carbon rods separated by an insulating layer made up mostly of kaolin clay. The carbon rods burned from the top down, the insulation melting progressively; alternating current ensured that the two rods burned at the same rate. When they were first installed in the Avenue de l'Opéra in Paris in 1877, they so outshone the gas lights that they were soon installed elsewhere in the streets of Paris. The candles were usually clustered in groups of four. Sixteen were installed in Billingsgate on 29 November 1878 by the Société Générale d'Electricité, and twenty

[18] See A. A. C. Swinton, *The Principles and Practice of Electric Lighting*, Longmans, London, 1884, p. 52.

[19] For a fuller account, see M. MacLaren, *The Rise of the Electrical Industry in the Nineteenth Century*, Princeton University Press, 1943, pp. 68–69.

on the Thames Embankment a week later.[20] A contemporary writer observed:[21]

> The electric candle invented by M. Paul Jablochkoff has done much to render electric illumination feasible. This candle, which has been much employed in Paris and is at present burning nightly on the Thames Embankment, is extremely simple and entirely without mechanism.

The arc light developed by Charles F. Brush was more complex but more efficient, and outlived the Jablochkoff candle. It employed a solenoid-operated clutch which kept the carbon rods the right distance apart; he also introduced a second set of carbons with automatic changeover which doubled the life between recarboning. The Brush lamp with double carbon had a life of about eight hours, where the Jablochkoff candle did well to last an hour and a half. Brush sold his first lamp to a Cincinnati doctor in 1877, and within the next few years installed arc lights in Wanamaker's store, New York; in Madison and Union Squares, New York; and in a small public lighting system in Cleveland, where he was working.

Raworth and Siemens

Thus, when Raworth joined Siemens in 1878, electric lighting had barely begun. The arc light was proven, though it was not a total substitute for the common gas light because of its power: the problem was to 'subdivide the electric light' as contemporaries put it and the arc could only be used where strong illumination was needed.

Nonetheless its use was growing. In late 1878, an increasing number of places employed arc lighting: the Times Printing Office; Woolwich Arsenal; Pullars Dye Works at Perth; St Enoch's Station, Glasgow; the London Bridge Terminus of the London, Brighton and South Coast Railway; Cammell and

[20] R. H. Parsons, *The Early Days of the Power Station Industry*, Cambridge University Press, 1939, Chapter 1.

[21] Swinton, *op. cit.*, p. 109.

Wilson's steelworks at Dronfield (between Sheffield and Chesterfield); the pit head of the Trafalgar Colliery in the Forest of Dean; and the Holborn Viaduct in London, though the last named abandoned it as too expensive six months later. On 13 October 1878, a football match was played at Bramall Lane, Sheffield, under four arc lights of eight thousand candlepower; the kickoff was at 7.30 pm, 'the players being seen almost as bright as noonday'.[22]

Siemens Brothers were in the forefront of all this activity, both as suppliers of dynamos and as installation engineers. In 1879 they arranged electric lighting for the British Museum and gave an impressive display at the Albert Hall:

> Undoubtedly the most effective, and perhaps the most important, of the electric lights exhibited at the Albert Hall on the 7th and 8th inst was the 'electric sunlight' of Messrs Siemens Brothers, which occupied the entire dome of this great building.[23]

As Siemens representative in Lancashire and Yorkshire, Raworth was responsible for many early electric light installations. In Appendix 1 are listed the installations with which Dorman and Smith, as successor to John S. Raworth, were associated. Some of these were in fact attributable to Raworth, including the store of Messrs Lewis's and Company in Market Street, Manchester. Although the incandescent lights were installed later, in 1882,[24] the arc lights appear to have been among the earliest ever to illuminate a store, being installed at about the same time as Brush's pioneer installation at Wanamaker's store in New York. A publication of 1881 noted:[25]

> We are informed that this firm (Messrs Lewis's) have had the electric light in their Market Street

[22] *The Electrician*, Vol. 1, 19 October 1878, p. 253.
[23] *The Electrician*, Vol. 2, 17 May 1879, p. 301.
[24] According to a letter by J.H.Holmes of Newcastle to the journal *Lightning*, Vol. 1, 19 May 1892.
[25] *The Electrician*, Vol. 6, 30 April 1881, p. 303.

> premises for the last three years. They have 33 lamps in all; some of them are in use from 8 to 12 hours per diem. One large 6,000 candle light is erected at a great height above the building and can be seen for several miles round.

If this information is accurate, Raworth must have been remarkably capable to adapt to the new medium of electricity after his earlier specialisation in the more mechanical aspects of engineering. It suggests that he was wide ranging in his scientific interests and that his time with the Hopkinsons had been well spent. In 1879, Raworth continued to promote Siemens business, as witness the following account:[26]

> Messrs Siemens system of electric lighting is now to be seen in the Victoria Arcade (Manchester). The continuous and alternating system are used. Five lamps, worked by the alternating system have been fixed in the arcade by Messrs Siemens' agent in the Lancashire and Yorkshire District (Mr J. S. Raworth) and their working is most satisfactory, the light obtained being extremely brilliant and devoid of the constant flickering which so frequently mars the effect of electric lighting. Apart from these five lamps there is a display of the 'continuous' system, by means of a light which hangs suspended from the glass roof and which has now been open to public inspection for several weeks. The six lamps altogether give an excellent light.

A further milestone in Lancashire for Siemens Brothers was the introduction of their lamps for municipal lighting by Blackpool Corporation, which announced its intention to install six arc lights in August 1879. The lights, which actually numbered eight, were set up early in October and were the first major

[26] *The Electrician*, Vol. 4, 29 November 1879, p. 14.

municipal system in England.[27] However, it is not certain that Raworth was directly associated with the installation as a Mr Andrews is said to have acted as Siemens representative. It may be that Raworth was busy elsewhere; or that the contract, originally estimated at £2,500, was such an important one that it merited a senior representative from London, for it is interesting that William Siemens himself gave personal attention to the work.

The success that electric lighting enjoyed and the publicity it gained caused great misgiving in the stock market about the value of shares in gas companies. As early as the mid-seventies over-optimistic claims for electric light had caused serious slumps in gas company share prices, followed by recoveries, in which fortunes were made and lost. The following account of the Siemens lights at Blackpool is taken from the *Journal of Gas Lighting*:[28]

> Blackpool aldermen and town councillors are very wise in their generation. The thing is a grand speculation, and pays. At this advanced season of the year Blackpool is usually empty of visitors. At present it is full to repletion with 70,000 to 100,000 visitors. Not a room, or even a sofa, is to be had in any of the hotels, from the 'Imperial' downwards, unless it has been secured a fortnight beforehand. The lodging-houses, too, are overflowing, and excursion trains from all parts of the country are bringing thousands of trippers into the town, returning them after midnight, each day of the fête, to their respective destinations. But what is the impression it produces upon the mind of the spectator in regard to its effect as a competitor of gas lighting? Simply this, that the latter rises

[27] 'Blackpool is the town in which the electric light was first used on any complete scale for public lighting', *The Electrician*, Vol. 6, 11 December 1880, p. 38. The arc lights were much brighter than Jablochkoff candles, and totalled 48,000 candlepower.

[28] This account was reproduced in *The Telegraphic Journal*, 15 October 1879, p. 537.

> higher and higher in esteem as a useful, easily-managed, and ever-present illuminator. Electricity may be the champagne light of fête days, but coal gas is the Bass's beer light of everyday life; and though the former is not to be despised on occasions, the latter is the wholesomer, not to mention the cheaper of the two, and what most sensible people will prefer to use. It is a noteworthy and curious circumstance that every trial of electric lighting, however successful, only serves to prove its utter general impracticability as a general illuminator, and the impossibility of its ever, in the slightest degree, competing with or interfering with the progress of gas lighting! If gas managers could only be permitted to take the trouble and go to the expense of organising a display of the kind witnessed last week at the northern seaport, nothing would more contribute to prove the truth of the views here expressed, and the utter folly and simplicity of gas shareholders who would timidly dispose of valuable property – about the most valuable in the country at the present hour – at a sacrifice, in fear of such a competitor.

Raworth, to judge from the number of other Siemens installations in Lancashire and Yorkshire, was extremely busy from 1878 onwards. He seems to have been too preoccupied with day to day business to patent any inventions, for from 1880 to 1883 inclusive he did not apply for a single one, whereas before and after this period scarcely a year passed without several applications being filed. As we see from the list of installations by Raworth and his successors, Dorman and Smith, the work was not restricted to Lancashire and Yorkshire, since there are installations mentioned in London, Birmingham, Cheshire, and Monmouth and Caernarvonshire. In 1881 a Siemens machine was used to drive a saw bench at Middlesbrough in Yorkshire, but lighting remained the main activity of Raworth's business and it was rendered still more important by the arrival of the incandescent electric lamp. The arc lamp was impressive but too powerful for normal domestic use; the Jablochkoff candle was limited by its short life. But in December 1878, Joseph Swan of Newcastle demonstrated his incandescent carbon lamp: independently in the United States, Thomas Edison constructed a carbon filament lamp which burned for forty-five hours continuously in October 1879.[29] The incandescent light, preserved from combustion by the vacuum in which it glowed, was a reality. It is ironic that the extract from the *Journal of Gas Lighting*, quoted above, should have been published just as Edison's lamp triumphed: for the 'subdivision of the electric light' meant that gas lighting would meet competition in every application. Edison hurriedly patented his discovery, whereas Swan had not because he considered his early experiments in 1860 had demonstrated the principle of the carbon filament, and he could not therefore claim its novelty in 1879. However, Swan did patent the crucial process of evacuating the lamp while incandescent so as to remove the last traces of oxygen. Rather than embark on lengthy and expensive litigation, the inventors joined forces and set up the Edison and Swan United Electric Light Company, with its 'Ediswan' trade mark.

Swan's early installations were in private houses but in October 1881 the new incandescent lamps were introduced at the Savoy Theatre, with over 1,200 lamps, most of them illuminating the stage. In the United States, Edison's first commercial installation was in the SS *Columbia* in 1880, and in the next two years 150 plants powering thirty thousand lamps were installed. Raworth naturally extended his own activities to include the installation of incandescent systems. Appendix 1, listing Dorman and Smith installations, shows clearly how the incandescent or 'glow' lamp was used more readily in offices, shops and private houses than arc lamps had been.

Raworth and the role of electrical contractors

Nobody can gain a proper appreciation of the electrical pioneers of this period without understanding the novelty and individual-

[29] *The Electrician*, Vol. 6, 21 January 1880, p. 175.

PLATE 3. An early lampholder; the loops from the lamp had to be clipped over the spring hooks to make a connection. The Edison-Swan lamp shown was the pioneer among electric light bulbs

PLATE 4. Typical Dorman and Smith domestic lightfittings of the 1890s

PLATE 5. Main switchboard at the Mansion House, London. Dorman and Smith manufactured it in 1893 and it remained in service until 1965. It now stands in the foyer of Atherton Works on permanent loan from the Science Museum, South Kensington

ity of electrical installations. In their infancy most innovations call forth new practitioners – new entry is easy, as economists put it. But likewise in these early times, failure and exit become common also as competition weeds out the weak members of the industry from the strong. The state of the electrical arts in the late seventies and early eighties, particularly in matters of installation, was crude. The original telegraph wires between Camden Town and Euston in 1837, mentioned earlier, had been encased in wood. Siemens had introduced subterranean cables with gutta percha insulation, but at the level of minor individual installations, practices were chaotic. A technical historian records:[30]

> Very little information has survived of these installations – during the seventies the cable makers were really only experimenting with the use of rubber and there were no recognised ways for protecting the wiring. Earth return was commonly used to save expense, even gas pipes being employed for this purpose. Switches if they existed at all were crude pieces of mechanism mounted on wood, and fuses were still to be invented.

At almost every turn there were difficulties. Apart from a few very specialised basic items, such as dynamos, arc mechanisms and to some extent, electric wire, the contractor had to find his own supplies and design his own system, treating each successive installation as a separate exercise in design. There were very few suppliers of materials for electrical contractors, let alone specialised fittings, though Verity's had been established in Covent Garden in 1875 as manufacturers, wholesalers and contractors of electrical equipment. Raworth himself recalled that in about 1878[31]

> There was in Mill Street, Ancoats, a factory devoted entirely to the production of dynamos,

[30] J. Mellanby, *A History of Electrical Wiring*, Macdonald, London, 1957, p. 22.

[31] Extract from his Presidential Address to the Northern Society of Electrical Engineers, 1898; Wilde was in fact a most notable pioneer, who invented but did not patent the dynamo a year before Siemens took out their English patent.

> arc lamps and projectors . . . run by Mr Henry Wilde.

However, for the most part, electrical contractors had to find their own materials, and it has been aptly said that 'every installation was a monument to its designer'.[32] Virtually every electrical system had to be installed as a complete entity. This involved not only the provision of lights, wiring and a dynamo, but also a source of power for the dynamo; every installation, in effect, had to have its own miniature power station. Usually power was provided for the dynamo by a steam engine or, if space was limited, by one of the new Otto gas engines, which were also being built by Crossley Brothers of Manchester. Thus Raworth's strong background in mechanical engineering design was invaluable in his new business. Yet he was more than just a competent engineer: he was a considerable innovator. Apart from all the lighting installations, (and the growth of his business in the 1880s attests their success) he was the first person to construct an electric hoist. A biographical sketch of Raworth notes:[33]

> For a time he ran a small central station, distributing current in its immediate neighbourhood. Most of the important electric installations erected in that city (*ie* Manchester) in those days were done under his superintendence.

In providing a central power source, Raworth anticipated a fundamental development in electricity supply – the development of large power stations which could exploit the great economies of scale of large generators and distribute cheap power throughout an extensive network. However, as we shall see presently, prospects for central power generation suffered a severe setback, albeit temporary, in the early eighties, so that Raworth's contribution in this area did not lead immediately to greater things. No catalogues survive to show exactly what Raworth manufactured. In May 1884 he applied for a patent for an electric meter, and the 'Raworth Detector' remained in

[32] Mellanby, *op. cit.*, p. 28.
[33] *Cassier's Magazine*, November–April, 1902–3, p. 679.

production for many years after he left the business, and indeed beyond his death. In June, he sought a patent with R. A. Smith for couplings for metal-cased insulated electrical conductors. A number of entries in the wages book for piece rates show that in 1884–5 lampholders, keys, brackets, pendant fittings, joints, fused sockets, yardarm reflectors and lamps were some of the simpler items manufactured by individual employees. It seems quite likely that much of the subsequent range of products of Dorman and Smith, which included switches, distribution boards, and lamp fittings, followed from the original activities of Raworth. His greatest contribution was in the provision of lighting for ships.

Shipboard lighting

Arc lights were used earlier in lighthouses than in ships. The Royal Navy was possibly the first user of shipboard arc lamps on HMS *Temeraire* and *Alexandria* in late 1877; they were also ordered for *Dreadnought*, and though a contemporary reference said their 'nature and purpose was not specified'[34] their main purpose seems to have been as searchlights to help counter night attacks by the torpedo boats that had been introduced in the seventies. The first civilian ship to use the light was Siemens Brothers famous cable layer, SS *Faraday*, which was described in a contemporary account:[35]

> SS *Faraday*, moored at Gravesend, was illuminated by the electric light she carries. A separate steam engine on deck drives the apparatus, the Siemens 'Dynamo Electric Light Apparatus', whose 6,000 candlepower lamp allows one to read ordinary writing at a quarter of a mile. At sea it is placed at the mast head.

This installation had a direct use in facilitating cable laying at night, rather than for general illumination. Likewise a German

[34] *The Electrical Review*, 1 September 1877, p. 210.
[35] *Ibid.*, 15 October 1878, p. 245.

ship fitted with an arc light at Bremen in 1878 used it principally to navigate the river.[36] The following year saw the first arc light installed for internal illumination on board ship. The British Electric Light Company fitted a temporary lamp in the saloon of the new PSNC steel ship *Mendoza*,[37] built by Napier and Sons of Glasgow in 1879, but it was not retained.

The first successful commercial application was in the Inman liner *City of Berlin* in December 1878. Raworth was the driving force behind the installation, though both Siemens and the Inman Line welcomed it. The steamship company decided that the light on the passenger deck would give them the opportunity to test travellers' reactions to the new system. Until then ships were illuminated by oil lamps or candles. Gas lighting would have been better but it was quite impracticable on board ship, so passengers had to suffer the murk and stench of oil lamps. They were especially objectionable in the small cabins where portholes often had to remain closed throughout the voyage: indeed, this problem was not to be solved until the use of incandescent lamps a little later.

Public reaction to the twenty arc lights was enthusiastic:[38]

> During her recent trip to New York . . . The passengers have addressed a note of congratulation to the company. The experiment of lighting the vessel by electricity has proved highly satisfactory. Not only was the saloon brilliantly illuminated but the steerage, usually the gloomiest part of the ship, was lighted up in all parts.

Thereafter, ships' lighting provided a growing part of Raworth's business, especially after the advent of incandescent lamps which could be installed in the cabins. There were repeat orders from the Inman Line and new business from many others. Entries in a wages book in later years suggest that Raworth was also responsible for maintenance work on the ships' lighting systems which he installed. It was not all plain sailing, and there

[36] *The Electrician*, Vol. 1, 21 September 1878.
[37] *Ibid.*, Vol. 3, 8 November 1879, p. 290.
[38] *Ibid.*, Vol. 4, 3 January 1880, p. 73.

was opposition. An electrocution on board a Russian ship with electric light brought warnings from the chairman of P & O and some installations were unsuccessful: the Orient liner *Chimborazo* was fitted with seven arc lamps[39] (not apparently by Raworth) but these were later removed and forty incandescent lamps installed by Raworth, with Killingworth Hedges, a noted electrical engineer, as a supervisor. The business relationship between Hedges and Raworth is not clear. Apparently, the Electric Lighting Supply Company was the main contractor with Hedges acting as supervisor, but Raworth did most of the actual design and work of installation. They collaborated in March 1881 in another Inman Line contract for the *City of Paris*, which was fitted with eight arc lights in the steerage.

An increasing number of ships adopted the new light, and though Raworth often used Siemens equipment he also used the new Swan lamps. In the Cunarder *Servia*, Raworth installed incandescent lamps in the engine room as well as the public rooms, powered by a Siemens dynamo.[40] We can gain some measure of the pace of business from the following contemporary account.[41]

> The SS *City of Berlin* and SS *Potosi* have been fitted with Siemens Brothers differential lamps. The SS *City of Paris*, the *City of Rome*, the Cunard SS *Servia* and Guion SS *Alaska* are all partly fitted with Swan lamps and partly with the differential lamps. Also the SS *Trojan* of the Union Company has one continuous lamp. The dynamo machines in all these instances have been made by Messrs Siemens Brothers and are of their well known types.

Raworth was associated with six of these seven ships, and therefore 876 out of a total of 877 lights. It required considerable expertise to overcome the problems of shipboard installation, and Raworth rose to the challenge. He was the first shipboard

[39] *The Electrician*, Vol. 5, 3 July 1880, pp. 73–4.
[40] *Ibid.*, Vol. 7, 6 August 1881, p. 178.
[41] *The Electrical Review*, 15 September 1881, p. 360.

contractor to install incandescent lamps in parallel, on board *City of Rome* in 1881, so that the failure of one bulb did not extinguish all the others; on this ship he also installed the first fusible cutout – obviously a very desirable precaution in a parallel circuit. Later, in 1884, he equipped the Guion liner *Arizona* with the first compound continuous current dynamo.[42] His basic work laid the foundations of ships' lighting practice for many years to come, and it was still possible to say at his death over thirty years later that his influence was reflected in every electric plant on board ship.[43] In an obituary we learn that:[44]

> There were difficulties connected with the system of generating the electricity. The early dynamos were belt driven, and the arrangement was not convenient in the confined spaces in ships, which precluded the use of a belt of sufficient length. Mr Raworth's first alternative was to use a continuous rope, upon pulleys with six or eight grooves, bringing the end back from the last driven pulley to the first groove of the driving pulley by means of a guide pulley. Later another arrangement was got out for a friction drive with the dynamo hung freely on two pivots so that the main bearings could adapt themselves exactly to the position of the spindle. These were more or less transitory stages. Later the high-speed engine directly coupled to the dynamo, of which Mr Raworth's 'Universal' engine was a notable example, overcame the difficulties, although it, in its turn, gave place to the turbine. But he had, in these early days, to devise also fittings, switches, switchboards, casings and the like, because it must be remembered that interior lighting was then in its infancy.

Such was John Raworth's contribution to shipboard lighting.

[42] *The Electrician*, 'Blue Book', 1904, p. XCII. An entry in Raworth's wages book shows that he had also done work on *Arizona* in November 1882.

[43] *The Electrical Times*, 29 March 1917, p. 241.

[44] *Engineering*, 30 March 1917, p. 308.

The consensus is that he did more than any other man to introduce and popularise it.

The progress of Raworth's business

Unfortunately, hardly anything of Raworth's business records survives. The only item in the possession of Dorman Smith Holdings today is a wages book, stamped 'John S. Raworth, agent for Siemens Brothers and Co Ltd' covering the period from mid-September 1881 to mid-March 1885. However, even this wages book is something of an enigma, for in it some people are recorded in a few periods, but not thereafter, even though we know for certain that they worked with Raworth until his departure to London in 1886. The young men who later bought the business, Dorman and Smith, appear only in February, March, July and August 1884; likewise R. A. Dawbarn, who was described as Raworth's 'right hand man in ship lighting work'[45] and who went with Raworth in 1886 to join Brush. Another man whose name disappears is G. Wilkinson who did well in the business since his wages rose from ten shillings a week in June 1883 to thirty shillings less than a year later; yet a George Wilkinson accompanied Raworth to the Brush Company in 1886.[46]

A possible explanation is that Raworth employed some people for Siemens and others on his own account, switching from one to the other as the work warranted. But it is more likely that Raworth simply acted as a supplier of Siemens equipment, and all installation and manufacture was his own business. In this case the surviving wages book probably covered only those employees engaged in the workshop and on outside contract work. The irregularity of hours for some of the employees suggests outside work, and so does the occasional payment of expenses. Other employees on flat rates were probably in the workshop, including a number of lads or apprentices paid five shillings a week. Dorman, Smith and Dawbarn, with their high

[45] See H. M. Sayers, 'The Brush', *The Electrical Times*, 19 May 1921, p. 480.
[46] *Ibid.*, 26 May 1921, p. 503.

rates of pay, would have been in the design and drafting office; both Dorman and Smith had served apprenticeships and Dorman had prior experience of design and took out a joint patent with Raworth in 1884. Dorman, Smith, and Dawbarn only appear in the wages book when activity is unusually high; it thus seems that they were called in to help production at critical times, for skilled labour was extremely scarce in the industry.

The lack of evidence makes it difficult to pinpoint exactly when Raworth began as an independent electrical engineer, and how old the business of Dorman and Smith may be said to be. Raworth was Siemens' agent from 1878 onwards, and not simply their employee. He installed electric lights soon after he became associated with Siemens. Given the embryonic state of electrical manufacturing at that time, and the fact that Raworth had earlier interests as an engineer and machinist, it is as near certain as undocumented fact can be that Raworth must have begun manufacturing some elements of his electric light installations in 1878; he could not simply have assembled complete kits sent to him from Siemens factory, because they did not exist. Thus the true foundation date for the present business of Dorman Smith Holdings would seem to be 1878, making it one of the oldest-established and longest-lived electrical companies in Britain and, indeed, the world.

In the absence of any other evidence, we have to try and glean from the wages book some indication of the success of the business. In 1881, there are three regular employees in the wages book, Staton, Stannier and Aldred, with respective wages of 22s, 18s and 15s for a forty-nine hour week. Thus the weekly wages bill shown was £2 15s 0d. From then on there was a steady growth in employment and in the total of wages paid. The latter gives some approximation to the growth of turnover, though probably not an exact one, as there are anomalies in the numbers of hours recorded: for example, during the second week of October 1881, all three employees were credited on two days with 29 hours, and 30¼ hours pay. Not even the energy and devotion to work of our Victorian forefathers can explain this away: in all probability they were paid double time to finish particularly urgent contracts, or possibly given some sort of incentive bonus. Thus the wages payments must be interpreted with a certain amount of care.

The general picture we gain from the wages book is one of rapid and steady growth. From employing four men in the first week of January 1882, the business expanded to employ thirty people in the corresponding week three years later, plus, presumably, Dorman, Smith and Dawbarn. The wage bill increased tenfold, that is, rather faster than employment, at a time when prices were falling. The table beneath shows the progress of employment and wages, as recorded in the wages book.

TABLE 1

RAWORTH'S WAGE PAYMENTS AND EMPLOYMENT, 1881–5

Year	January			April		
	WAGES		EMPLOYEES	WAGES		EMPLOYEES
	£	s		£	s	
1881						
1882	3	12	4	7	6	6
1883	15	10	15	18	18	17
1884	22	18	25	28	3	25
1885	36	6	30			
	July			**October**		
	WAGES		EMPLOYEES	WAGES		EMPLOYEES
	£	s		£	s	
1881				2	15	3
1882	8	8	8	7	10	8
1883	20	3	23	17	8	19
1884	23	0	25	29	13	26
1885						

Note: Figures are for first full week of month shown. Wages are rounded to the nearest shilling.

When Raworth first gained the Siemens agency, his business address was shown in trade directories as Exchange Buildings, Saint Mary's Gate, Manchester. In 1884 the expanding business moved a quarter of a mile to 22 Brazennose Street, where it was described as an electric light company. Next door at number 24, as agent for Siemens, Raworth shared the building with three other tenants. However, in 1885 he apparently concentrated his interests under one roof at number 24 (*Plate 2*). The building still stands eighty-seven years later, as a club.

The growth of employment in the business suggests strongly that profits were growing; and there is circumstantial corroboration from Raworth's domestic changes, for in 1882 he moved from 16 Brunswick Street, near the University, to the then greenery of Heaton Chapel.

One may wonder, therefore, why Raworth chose to abandon this apparent prosperity, pack his bags and go to work in London. We may speculate about the fortunes of the business, as there is no wages book for the period from March 1885 to March 1886, and that was a severe period for business generally. We may speculate about his ambition, for ambitious he certainly was. He had been a pioneer in his field and was well acquainted with leading figures in the industry such as Hopkinson, Killingworth Hedges and Robert Hammond.[47] In the early years of the decade, Raworth's commercial horizons must have seemed almost limitless, no less than the future of the electric light itself. Siemens was the leading company in electric lighting; the incandescent lamp vastly extended the scope of electric lighting and the economy was strong, having recovered from the depression of the 1870s; and Raworth was an acknowledged expert in his field, with a growing reputation which extended far beyond Manchester.

[47] His exact connection with Hammond is obscure in the early days. Hammond was the first concessionaire of the Brush company in England, and was extremely active in installations in the early eighties. In a meeting of the Northern Society of Electrical Engineers, Mr Henry McLaren said Raworth 'first started and stumped the country with Mr Hammond (who) put him in the position of providing the power'. *Proceedings of the Northern Society of Electrical Engineers*, Vol. 3, 1896.

Disappointments of the mid-1880s

Despite this golden combination of circumstance, disappointments followed. Siemens met increasing competition and decided shortly afterwards to withdraw from their unprofitable lighting activities and concentrate on their profitable submarine cable business.[48] The economic situation worsened, and after 1883 another depression developed from which there was no sign of recovery until 1886, eventually provoking serious riots in London and Manchester, and, according to a noted historian of the period, it 'gave Victorian optimism and courage the severest shock it had yet received.'[49] By 1886 tramp freight rates had fallen to an extremely low level and steamship construction, which had boomed over the years from 1876 to 1883 and provided Raworth with a great deal of business, plummeted in 1886 to less than a third of the 1883 level. Unemployment in the engineering and shipbuilding trades rose above the level of the depressed 1870s. The statistics listed in Table 2 show some of the influences which helped Raworth's business in the early eighties, but brought disappointment in the middle years of the decade.

But the biggest disappointment, for the whole industry, was the Electric Lighting Act of 1882, promulgated and passed by Joseph Chamberlain, the President of the Board of Trade. The government attitude had been affected by the monopolistic abuse of gas and water supply companies in the past, and by the Brush boom. Floated in late 1880, the second Anglo-American Brush Electric Light Corporation Limited had been received by an enthusiastic and gullible public. Brush had a sound technical reputation and sold exclusive rights to use its equipment to various local companies, and from their proceeds paid an imprudent hundred per cent dividend in 1881. Siemens Brothers, by now a public company, paid an eighty per cent dividend in the same year, though on a sounder footing. A speculative boom

[48] See J. D. Scott, *Siemens Brothers, 1858–1958*, Weidenfeld and Nicolson, London, 1958, pp. 64–5.

[49] Sir Robert Ensor, *England 1870–1914*, Oxford University Press, 1936, p. 111

TABLE 2 ECONOMIC INDICATORS, 1878–87

Year	Index of tramp freight rates (*1869* = 100)	Steamships built and first registered (*thousands tons*)	Percentage unemployed among members of engineering, metal and shipbuilding trades unions
1878	91	499	9.0
1879	85	412	15.3
1880	87	474	6.7
1881	87	486	3.8
1882	81	610	2.3
1883	75	806	2.7
1884	64	570	10.8
1885	63	393	12.9
1886	59	308	13.5
1887	65	322	10.4

Source: B. R. Mitchell and P. Deane, *Abstract of British Historical Statistics*, Cambridge University Press, 1962, pp. 64, 221 and 224.

followed,[50] but the bubble was pricked by the Act of 1882. The Act gave local authorities the right of compulsory purchase of electric lighting companies after twenty-one years on the base value of plant and machinery, with no allowance for goodwill. It made public lighting almost a dead letter, as the statistics of applications for public lighting show:

1883; 106 applications for Provisional Orders, of which 69 were granted
1884; 4 applications, 4 granted
1885; 1 application, refused

Further evidence of the inhibiting influence of the legislation is provided by the pattern of company registrations. The nominal

[50] It prompted misgivings in several quarters. See 'Electric Light and Power mania', *The Economist*, 20 May 1882, pp. 604–5.

capital of companies registered is a measure of the optimism of their promoters. Table 3 shows the total collapse of hope for electricity supply companies in the mid-eighties and the resultant loss of confidence in the electrical manufacturing industry, which catered for the derived demand for electrical lighting equipment.

TABLE 3 NOMINAL CAPITAL OF ELECTRICAL COMPANIES REGISTERED 1879–86

Year	Electricity Supply Companies (£)	Electrical Manufacturing Companies (£)
1879	75,000	55,000
1880	—	1,200,000
1881	1,079,000	65,000
1882	12,603,000	8,720,000
1883	2,140,000	1,739,000
1884	526,500	233,000
1885	58,000	790,000
1886	234,000	1,656,500

Source: Garcké's *Manuals of Electrical Undertakings.*

For the electrical industry, the legislation was disastrous. In 1882 it stood the equal of foreign competition but after six years of such powerful inhibition there were only seven electricity supply works operating, and the American and German industries gained valuable trade by the time the Act was amended in 1888.[51] In effect the Act meant that electrical installations were only made where the customer could afford his own generating plant, as in railway stations, large factories, department stores, hotels and large mansions and, of course, ships. The shipping business did much to help Raworth but he was undoubtedly irritated by the legislation and he would have met Emil Garcké, secretary of the reorganised Brush company, who played a leading part in the agitation that led to amendment of the Act. Later the two were to become long-standing business associates.

Of themselves, all these irritations might not seem enough to sway Raworth's choice; but it is possible that the business suffered seriously in 1885–6. There is, unfortunately, no wages book surviving for this period, but in Chapter Two we note the much reduced employment in the new business under Dorman and Smith in April 1886. Raworth certainly took some people with him to his new employer, Brush, but hardly enough to cause such a contraction. It therefore seems likely that 1885–6 was a disappointing year in Raworth's business. An attractive offer from Brush and a ready sale for his business would have been a strong temptation and he joined Brush as superintending engineer in April 1886.

There is one other potential factor. There was much correspondence in the electrical journals of the time about the relative costs of gas and electric lighting. Much of it neglected capital cost, and therefore overstated electricity's case because usually electricity had to justify itself against an *existing* gas installation. Byatt has conducted a detailed investigation of the period and concluded that in the early eighties, electricity was too expensive to compete with gas.[51] In special circumstances such as shipboard lighting, electricity had the edge, but elsewhere its success rested upon bigger and more efficient generators and distribution systems. There was little scope for Raworth to bring about such changes in his relatively small business in Manchester. Yet it is not altogether fanciful to suppose that he might have had ambitions of this sort, given the necessary resources; for one of his important achievements when he did join Brush was to design and build 1,000 hp generator sets, where previously they had made nothing beyond 50 hp dynamos.

Epilogue

Raworth was very successful at Brush. We cannot here examine the whole of his later career but only a few of the highlights. He

[51] I. C. R. Byatt, *The British Electrical Industry, 1875–1914*, unpublished D.Phil. thesis, Oxford, 1962, pp. 37–42.

used his contacts among the steamship companies to gain entry for Brush into this market for electrical installations. Under his direction, power stations were established in London, Leicester, Huddersfield, Hanley, Bournemouth and many other towns. He organised and equipped the new Brush works at Loughborough. Raworth went with Garcké in 1888 to Temesvar, in Austro-Hungary,[52] to reorganise its ailing electricity system. The enterprise became highly successful, driving the rival gas company out of business and eventually being bought up by the municipality. He became joint manager of Brush and a director, and by 1899 was on the boards of twenty-eight electrical companies, mostly in supply and traction and connected with Brush.[53]

Despite all these commercial preoccupations Raworth remained an enthusiastic engineer and innovator. Between 1886 and 1910 he applied for no less than eighty-three patents – twice as many as his protégés, Dorman and Smith. Most of his work concerned steam engines and electric traction, though in both these full success avoided him. He designed his 'Universal' high speed steam engine as a compact and efficient dynamo drive in 1896. It was used and was a genuine advance, but was overshadowed by the development of the Parsons turbine. He made important contributions to electric traction and started his own company, Raworth's Traction Patents, with offices in London and Manchester, but his regenerative system of control failed to be universally adopted. But

> if it failed of general adoption and commercial success, (it) showed great ingenuity and deserved a better fate.[54]

However, he nonetheless played an important role, including many years as the technical director of the British Electric Traction Company and, in the opinion of some, displayed his greatest technical ability in this field:[55]

[52] Now Timisoara in Romania.

[53] For the companies concerned see *Garcké's Manual of Electrical Undertakings*, Vol. 4, 1899–1900, p. 945.

[54] *The Electrical Times*, Vol. 51, 12 April 1917, p. 269.

[55] *Engineering*, 30 March 1917, p. 308.

> ... the result was the application of electric tramways through the enterprise of Mr Raworth and his colleagues to many small communities long before the larger cities enjoyed this advantage, because the corporations of those cities would not themselves act by extending the leases of those companies to make it worth their while to adopt electric traction.

He was an active member of the Institution of Electrical Engineers and the Institution of Civil Engineers and a very frequent contributor to the electrical engineering journals.

Raworth's character and influence

We do not know how much influence Raworth had upon the two young men who bought his Manchester business from him in 1886; his character was so strong that it must have left its mark. In his work, he was enthusiastic and prolific:[56]

> J.S.R. was a man who took a pride in his work. It had to be just right technically and artistically. He had been trained in the Lancashire textile machinery works, and like most graduates from that excellent school of engineering, his mind was replete with mechanical notions and dodges. He would sketch out half-a-dozen different ways of doing a job whilst the assistant or draughtsman who was trying to take his instructions was assimilating the first one. But, fortunately, he always decided himself which was the best among so many. As he had a fine artistic taste as well as excellent judgment in the practicability of ideas, the results were good.

Equally, he was a man of firm views about economic and social affairs. He was deservedly critical of restrictive legislation, such as the Electric Lighting Acts, the Road Locomotion Act,

[56] *The Electrical Times*, Vol. 51, 12 April, 1917, p. 269.

PLATE 6. The Ordsal Works, December 1891; drawing based on an illustration from a company catalogue

PLATE 7. The Ordsal Works later in the 1890s

the Tramway Act and the Light Railways Act, and believed in price cutting to encourage electrical demand, thus:[57]

> I know one electric light company which . . . reduced their price 40 per cent. This reduction involved them in an absolute loss, but as the property was owned by the directors they did not fear it – they had more faith in logic than high prices. The result was *one* adverse balance sheet, and then a mighty wave of prosperity.

yet he had a proper appreciation of real costs:[58]

> In comparing companies with municipalities, interest on capital ought to be charged to cost as well as depreciation, because a municipality has to pay interest and sinking fund before any profit appears. Shareholders have been educated to look upon everything they receive in the way of dividend as profit, whereas in reality only the surplus beyond the revenue obtainable from Consols is profit.

In debate and discussion, which he obviously relished, he was respected and impressive. He was always courteous and accessible to his staff and ready to help them with suggestions, and was as keenly interested in managing men as he was in engineering, though he was always anxious to stress that man was not a machine, and achieved his highest development in intelligent independence.

One biography describes him:[59]

> . . . with a ready wit, particularly fertile in illustrations that strike home; with oratorical powers of a high order, and with a manner that generally conciliates and always commands respect, he is an advocate very difficult to withstand, either in

[57] *Lightning*, Vol. 14, 25 August 1898, p. 159.
[58] *Ibid.*, 8 September 1898, p. 203.
[59] *Cassier's Magazine*, loc. cit.

> public or in private. At technical societies his rising is met with a hush of expectancy. At shareholders' meetings he forms an ideal chairman, securing confidence and disarming opposition by his frank dealing and his evident knowledge. At a public dinner, the air is rent with laughter as long as he is on his feet. In a word, he has the gift of securing attention to whatever he wishes to say, and the power of adapting his methods to the circumstances in which he finds himself.

Raworth kept in contact with Dorman and Smith, becoming a member of the Northern Society of Electrical Engineers in November 1895, and travelling from London to attend the meetings, to some of which he gave lectures. In 1898, he succeeded Charles Dorman as president of the society. There was clearly great mutual respect, for they were men of a kind: extrovert, able engineers yet possessing sound commercial sense. Raworth was a pragmatic, continuous inventor, not a revolutionary who made a single outstanding contribution. He never achieved the status of Ferranti or Crompton, and his name is nowhere commemorated. Yet his contribution was real and lasting, and in one form, as Dorman and Smith, his work was carried on.

CHAPTER TWO

The Partnership 1886–1914

In 1886 J. S. Raworth left Manchester to join the Brush Company in London. His business was sold to two of his employees, Charles Mark Dorman and Reginald Arthur Smith, who established a partnership which began business on Lady Day, 25 March 1886. The partnership was styled 'Dorman and Smith, Electrical and General Engineers and Electricians, manufacturers of plant and fittings'. It advertised as 'Successors to John S. Raworth' and took over his premises at 24 Brazennose Street.

According to the *Electrical Review* for 1920 both Dorman and Smith became assistants to Raworth during 1881. However, the first concrete evidence we have of their employment by Raworth relates to 1884. In that year both their addresses are listed as c/o J. S. Raworth, 24 Brazennose Street, in the *Manchester Trade Directory* and in the *Journal of the Institute of Telegraph Engineers*. In 1884 also a patent is jointly taken out in the name of Raworth and Smith for electrical couplings,[1] indicating a close technical association. The decisive evidence, however, is to be found in Raworth's Siemens agency wages book, which includes certain wage payments to both these men, commencing at the beginning of 1884.

The partners

Charles Mark Dorman (*Plate 8*) was born at Northampton on 13 February 1861. He is believed to be the son of Mark Dorman, JP, of Melbourne Crescent, Northampton, who, as early as 1871, was a member of the Society of Telegraph Engineers, the

[1] Patent No. 9104 (1884).

forerunner of the Institution of Electrical Engineers. He was educated at Berkhamsted and the Yorkshire College. Afterwards he served an apprenticeship in the works and drawing office of the Hunslet Engine Company where, according to the *Electrical Review*,[2] 'he did much valuable work in the way of locomotive design'.

Reginald Arthur Smith (*Plate 9*) was born at Stoney Middleton in Derbyshire on 26 January 1857. He was the son of the Reverend Urban Smith (1804–87) who obtained a very distinguished classics degree at Trinity College, Cambridge, in 1830 and became the Rector of Stoney Middleton, near Sheffield, in 1834[3] and remained there until he died fifty-three years later. After leaving school, R. A. Smith served an apprenticeship at the Horwich locomotive works. Whether, during their apprenticeships, these two young men gained any contact with the early applications of electricity is not known. This, however, is a possibility since one of the first commercial applications of electric power was in telegraphy, which was soon incorporated into railway communications systems. Smith later went to work at the Siemens factory in Woolwich, though we do not know whether he was there at the same time as Raworth's brother, Alfred. He had an extensive range of friends in industry: when he applied for full membership of the Institution of Mechanical Engineers in November 1885, he was supported by William Denny, the Dumbarton shipbuilder who pioneered steel instead of iron in steamship construction; Charles Sacré, a veteran railway engineer of the Manchester, Sheffield and Lincolnshire Railway Company; Thomas Waterhouse, director of an engineering company in Sheffield; Robert Wyllie, the general manager of Thomas Richardson and Sons, marine engineers at Hartlepool; and R. H. Heenan, the owner of a Manchester engineering factory.

Whilst the two partners shared a common professional training, they were quite different in personality and physical appearance. Dorman was a large and jovial man who knew, and used, the Christian name of each of his employees. He had a keen sense of

[2] *Electrical Review*, Vol. 86, 20 February 1920, p. 244.
[3] Crockford's Clerical Directory, 1860.

humour and his laughter was of such resonance and volume that it would be heard through the greater part of the works. He was an active member of scientific societies in the north-west and also established a reputation as 'a good after-dinner speaker'.[4] He dressed comfortably rather than precisely, limiting his wardrobe to a couple of suits.

Smith was also tall but thinner featured. He was far more reserved and serious than Dorman. He spoke to his employees far less frequently than his partner and always in more formal terms. Though a member of certain scientific societies he spoke less often. Even their style of dress reflected the differences in personality. Reginald Smith was always meticulously dressed and maintained an extremely large wardrobe. According to the recollections of one of his employees, he never appeared in the works on two consecutive days wearing the same suit. This interest in clothes was retained after he retired from active management when he lived in Windermere in the Lake District, and when he attended board meetings in Manchester he would often call on his tailor there.

The outlook

In April 1886 Dorman and Smith must have been viewed as two young but capable men who had just taken a very considerable (though, no doubt, calculated) business gamble. They were entering an industry in its infancy in which the product had not been technically perfected and to which the public had already demonstrated a considerable sales resistance. Added to this, the 1882 Act had severely curtailed development and forced a number of the less fortunate pioneers into bankruptcy. Where did their future lie and what lines might they profitably follow? To answer this question it is necessary to look briefly at the structure of the embryonic industry as it existed in the mid-1880s.

Its first characteristic feature is typical of most industries in the early stages of development – the absence of any high degree of

[4] *Lightning*, Vol. 9, 30 January 1896, p. 82.

specialisation either horizontally (in the range of electrical products produced) or vertically (in the number of stages of the production process in which firms engage). The typical contractor for an electric lighting installation, though he would often need to purchase the power generator and cable from other manufacturers, would normally use his own engineering skill to design and produce many of the small electrical parts he required as well as undertake the installation work himself. This type of work attracted relatively small firms with a limited amount of capital. There were already, however, a number of much larger firms in the industry, especially in the field of dynamo production, cables and electric lamp bulbs. In dynamo production the market leaders were Brush, Siemens, Crompton, and Mather and Platt. In cables, the older established telegraphic firms were dominant – Siemens, Telegraphic Construction and Maintenance Company, India Rubber and Gutta Percha Company. In lamps, Ediswan was the major firm. Even by the mid-1880s, therefore, the industry was dividing into a small number of large firms producing the larger electrical products and a considerable and growing number of relatively small firms mainly producing smaller electrical products and engaging in installation work.

Raworth, in selling his own business and joining Brush, had clearly decided that he was unable to enter the heavy end of the industry on his own account. A similar problem faced the young partners because in 1886 they possessed neither sufficient capital nor technical knowledge to enter the heavy electrical section. During the 1890s technological advance led to larger generating units and the gulf between the large and small sized firm increased. Usually this could only be bridged by acquisition or merger; thus Dorman and Smith were committed to the lighter end of the industry, certainly so long as they retained the partnership form of ownership.

Within the lighter end of the industry their success depended essentially on three things; technical inventiveness, good commercial practice and a change in the statutory provisions relating to electricity supply. The first two of these factors may be regarded as the basic determinants of the firm's fortunes not only in the pre-1914 period but in subsequent years as well.

There were two basic technical requirements – to develop electrical parts that would suit particular installations and to redesign and develop parts to take account of the changes taking place in the generation and distribution of electricity. In the longer term, as the industry came to specialise, the partners would need to identify an area of manufacturing where they would be among the technical leaders. However, substantial technical innovation was not a sufficient guarantee of business success. During the first twenty years of the industry's development there were many firms who lacked what contemporary writers termed a 'commercial engineer' or 'engineer trader' – that is, someone who possessed a clear appreciation of market requirements as well as the ability to design the appropriate equipment. The early success of Dorman and Smith was to depend as much upon the commercial sense as on the technical capabilities of these two quite different personalities.

The early years

In the earliest days of the partnership Dorman and Smith continued to emphasise their installation work, building upon the reputation and contacts established by Raworth. One of their first publications (reproduced as Appendix 1) lists the installations of glow and arc lamps for which the firm and its predecessor were responsible and advising that they engaged in 'Complete Installations of Arc and Incandescent Lamps for Houses, Ships, Works, Mills, Warehouses, Shops, Offices, Theatres, Gardens, etcetera'. The list is dominated by thirty-six steamship installations reflecting Raworth's close association with the earliest electrical lighting installations on ships. This connection between the firm and the shipbuilding industry has continued to the present day though, except for war time periods, its relative importance has considerably diminished.

The remaining installations in the list include a variety of commercial and manufacturing premises (for example, two of Lewis's shops in Manchester and Liverpool, the Tate Sugar

Refinery and the Southport Winter Gardens), but only two private houses. Despite the restrictions on the public supply of electricity imposed by the 1882 Act, Dorman and Smith believed the private house market possessed considerable potential and devoted part of their advertising effort in this direction. Significantly, they saw their first task as convincing potential purchasers that electric light was a superior form of illumination for private houses to the alternatives which were available. This is illustrated by the following extract from one of their earliest circular letters (reproduced in full as Appendix 2):

> The light itself is as perfect as any artificial light can be. It is steady, soft, and of any desired brilliancy. It gives off no fumes, and very little heat; hence it may be used to any extent, without compunction, in drawing rooms, etc, without fear of damaging painting, flowers, or the most delicate fabrics, and, what is of much more importance, without fear of injuring the health of the occupants. It is very handy. This is especially the case when the dynamo is supplemented by an accumulator, in which case a light can be obtained in any part of the house, at any time of the day or night, by simply touching a switch, which can be placed in any convenient position – for instance, near the door in a dwelling-room or close to the bed in a bedroom. It gives almost absolute safety from fire, abolishing the risk of explosion, inseparable from the use of gas, and the danger of overturned lamps and candles, which has been the cause of so many destructive fires.

This same circular letter incidentally sheds light on another interesting question – why Dorman and Smith did not continue the Siemens agency which had been held by Raworth:

> (We) are in a position to obtain engines, dynamos, etc, from all the best makers upon very favourable terms, not being tied down to any particular

> firm, we shall, in every case, be able to make the selection which our experience shows us to be the best.

Not surprisingly, we find Dorman and Smith acquiring such equipment from Brush as well as Siemens!

This early emphasis on installation work was, however, shortlived. By 1887 or, at the latest, 1888[5] they had given up installation work and had come to specialise exclusively in the manufacture of electrical parts. Nevertheless their reputation for installation work lived on and they found it necessary, in their 1891 Catalogue ('to save useless enquiries') to state explicitly that such work was no longer undertaken. 'The electrical industry is now too large to allow of any one firm grappling successfully with all its sections' – the industry had developed to a point where some further specialisation was necessary.

The reason for this early termination of installation work by Dorman and Smith lay in their initial technical successes which, within a matter of months of the partnership being founded, determined the area of electrical part manufacture in which they would specialise. In May 1886 a patent was taken out in the name of R. A. Smith for electric lampholders and, in June 1886, another patent was recorded in the name of C. M. Dorman for electric switches. Three other patents for lampholders and switches were registered in their joint names later in 1886 and during 1887.[6] According to the *Electrical Review*, Dorman and Smith displayed their products at the exhibitions held in Liverpool (1886) and Newcastle upon Tyne (1887) and on both occasions won medals. The basis for their reputation as specialist manufacturers of switchgear and ancillary lighting equipment was established.

Both the two initial patents, Dorman's switch and Smith's ceiling rose, were important in the immediate success of the

[5] According to the *Electrical Review*, Vol. 61, 5 July 1907, pp. 19–21, they had undertaken no installation work during the last twenty years, but this may be a slight exaggeration.

[6] Between 1886 and 1912, Dorman and Smith and their associates applied for forty-one patents,

company, though in the long run switchgear predominated. The term 'switch' dates back to the early days of telegraphy. In railway parlance, it was the American equivalent of 'points' which switched trains from one line to another. In the electrical engineer's vocabulary it initially referred to a device for switching the current from one circuit to another; later it was applied to any device for interrupting the current. There were both main and branch switches, the latter made necessary by the introduction of the incandescent lamp. The earliest branch switches which were used in the first half of the 1880s were largely influenced by the mechanism of the gas tap; they had a turning device with no definite on or off position and their base and cover were generally made of wood. As such they were not sufficiently insulated or safe from fire.

The developments in branch switches during the later 1880s were of three kinds: the substitution of non-combustible materials, the development of definite on and off positions and, later, quick make and break mechanisms. Dorman and Smith were among the pioneers in each of these three developments but the main significance of Dorman's switch is that it appears to be the first to have a porcelain base, a practice that was to be followed almost universally for many decades. Moreover, the patent lampholder also contained a porcelain nozzle covering the terminals, indicating that their use of porcelain was not limited to the switch.

Nor did their technical inventiveness end with the successes of 1886 and 1887. In 1888 they had no less than eight patents approved. From that year, through to 1914, the partnership obtained at least one new patent, and sometimes two or three, in virtually every year. In the period up to 1893 these included further patents for lampholders and other ancillary fittings. After this date, however, they almost exclusively related to switches, fuses and cutouts. In 1892 they followed up their earlier successes at Liverpool and Newcastle by gaining a medal for their products at the Crystal Palace exhibition, and winning the contract for the Mansion House switchboard in 1893 (*Plate 5*).

Their interest in branch switches soon extended to main switches and switchboards. The main switch controlled the supply of electricity to a whole building and it needed to satisfy certain important requirements. It had to make good contact, be unaffected by vibration and, whilst remaining in good condition untouched in a 'closed' position, yet be capable of operation at any time under load without suffering damage. The early developments of main switches took two main forms – 'brush contact' and the 'knife' – and there was considerable contemporary argument as to which was the superior method. In the former method the edges of springy contacts of phosphor bronze or similar material bore on a flat metal surface whereas the 'knife' took the form of a contact blade. Dorman and Smith opted for the second of these two methods in their patented main switch and, in accordance with early practice, mounted their main switches on slate panels without any enclosure.

From their catalogue of 1891 (which is the earliest known to exist) it is possible to obtain a clear picture of the influence of these early technical advances on the product range they manufactured and distributed. The main emphasis in the catalogue is placed on switches, switchboards, cutouts, lampholders and ceiling roses (*Plates 3 and 4*). Most of these were patented by Dorman and Smith and all of the smaller items had the characteristic porcelain component. In addition they supplied a very wide range of small lighting accessories together with certain special lightfittings for ships and also Raworth's popular Lineman's Detector.

The technical basis for the partnership's success as a specialist manufacturer of switchgear and ancillary electrical equipment was therefore established in the very first years of its existence but its market potential was curtailed so long as the restrictive provisions of the 1882 Act remained in force. Then, in 1888, an amending Act was passed which extended the security of tenure for enterprises providing a public supply of electricity from twenty-one years to forty-two years. The resulting boom in the establishment of central generating stations, especially in London, inevitably meant a sharp increase in demand for all forms of electrical equipment with a continuing improvement in the long term prospects as the reliability of supply improved and the price of electricity fell.

Dorman and Smith responded to this development by establishing a London office, towards the end of 1889, at 11 Grocers' Hall Court. Their annual profits in both 1889 and 1890 were ten times greater than in 1886 whilst the annual wages bill had doubled in the intervening period. The rate of expansion in orders was almost an embarrassment because it placed acute pressure upon their limited working accommodation. They possessed a small works in Hulme Street, adjoining their offices in Brazennose Street, but this could not be satisfactorily expanded to meet their increased requirements. They were therefore confronted with the major decision of acquiring new and greatly enlarged working premises.

The move to Salford

This problem was solved by the purchase of a new factory on a site in Middlewood Street, Salford, to which the firm moved on 25 March 1892. This new factory, Ordsal Electrical Works, was proudly described by the partners as 'the most complete for our branch of the industry in the country'.[7] (*Plate 6*) Although the area was called Ordsall, Dorman and Smith called the factory the Ordsal Works, to emphasise their initials DS exactly in the middle of the name, thus OrDSal.

The move was shortly followed by a number of important management changes. After superintending the move to Salford, A. J. Hallam, the works manager, retired at Christmas 1892 and was presented with a gratuity of five pounds by the firm. Hallam had been an employee of J. S. Raworth and his name was first recorded in the wages book at the beginning of 1884 (at the same time as both Dorman and Smith). His replacement as works manager was an outside appointment, Herbert George Baggs. From 1894 onwards Baggs is associated with most of the patents that were taken out in the name of the partnership and he was later to rise to the position of chairman. A detailed study of him and his work will, however, be reserved to the next chapter. Later, in 1893, the London office was placed under the control of

[7] Foreword to the 1891 catalogue (see Appendix 3).

A. H. Dorman, the partner's brother. Further attempts to strengthen the marketing arrangements were taken in 1895 when T. A. Nunwick, formerly branch manager of Poole and White Limited, was appointed Dorman and Smith representative for the northern and midland counties[8] and an office was opened in Glasgow, managed by Andrews and Company.

The prospects of the partnership, at the time of their move to Salford, must have appeared very bright and, by the end of the first year in their new location, they recorded that the move to larger premises had been entirely justified.

> Even with increased facilities, their heavy department has been overtaxed and the men kept on overtime a great part of the year. In smaller gear, such as small switches, fuses, fittings, etc, while there has not been such a marked increase in the orders given out by the leading contracting houses, the amount of material used by the smaller wiring firms, and the enormous increase in the number of the latter during the year shows that there is a tendency, especially in the provinces, to employ local men for wiring contracts. Firms engaged on ship-lighting have apparently not been quite as busy as usual, especially towards the close of the year. The large number of central stations now in progress up and down the country are gradually but surely increasing the proportion of goods used in the provinces and it is in this direction, rather than in London, that the firm looks for great extension of output in the coming year.

The report ends on a note of optimism which also reveals the emphasis they placed on the quality of the products they produced:

> Messrs Dorman and Smith are very pleased to record the fact that there is a distinct tendency to use

[8] *Lightning*, Vol. 7, 28 March 1895, p. 210.

> better material, and that they do not now feel the unhealthy competition of cheap English and foreign rubbish at all.[9]

This degree of optimism, however, was not entirely justified. Whilst total sales employment and gross profits on goods were increasing slightly at the time of, and in the years immediately following, the move to Salford, the same was certainly not the case with net profits. Net profits in each of the two years immediately following the move were only fifty per cent of the level achieved in 1890 and, in view of the considerable increase in fixed assets, this caused a very noticeable deterioration in the capital/earnings ratio. There were two reasons for this drop in profit – a disproportionate increase in overhead charges (notably rents and depreciation charges) and in wage and salary payments and, secondly, a deterioration in the general economic conditions in the trade.

General economic indicators for the electrical trade in this period are very sparse and not entirely satisfactory. However, in the Table below, the net profit record of the partnership, in the period 1886–1913, is compared with two available indicators – unemployment rates in engineering, metal and ship building unions, and new steamship registrations by tonnage. In 1886 and 1887 unemployment rates were high whilst ship registrations were at a low level. The period 1888–90 saw a substantial improvement in both indicators, partly because of the higher level of shipbuilding activity and the favourable impact of the 1888 Act. This is the period in which Dorman and Smith's profits increased sharply and the need for new premises became evident. However, the initial boom was not sustained and both indicators show a considerable decline in the period 1891–3, extending to at least the end of 1894 in the case of the unemployment indicator. This, in turn, coincides with the sharp downturn in Dorman and Smith's net profit performance. The extent to which the deterioration in performance was due to a temporary setback in general trade conditions or to more fundamental long term factors will be examined later in the chapter.

[9] *Electrical Engineer*, Vol. 11, 6 January 1893, p. 19.

TABLE 4

DORMAN AND SMITH NET PROFITS AND SELECTED ECONOMIC INDICATORS, 1886–1913

Year ending	Net profits (£)	Percentage unemployed in engineering, metal and shipbuilding unions	Steamships first registered (*thousands tons*)
1886	119 (9 months)	(13.5)	155
1887	684	(10.4)	225
1888	911	(5.5) 6.0	407
1889	2,015	2.3	554
1890	2,304	2.2	529
1891		4.1	479
1892	2,165 (18 months ending June)	7.7	434
1893	1,204 (June)	11.4	380
1894	1,281 (June)	11.2	486
1895		8.2	466
1896	4,860 (21 months ending March)	4.2	463
1897		4.8	416
1898		4.0	654
1899		2.4	704
1900		2.6	698
1901	13,984 (5 years ending March)	3.8	721
1902		5.5	736
1903	5,434 (2 years ending March)	6.6	587
1904	2,907 (March)	8.4	702
1905	2,183 („)	6.6	821
1906	3,204 („)	4.1	890
1907	2,778 („)	4.9	717
1908	3,262 („)	12.5	387

1909	1,720 („)	13.0	484
1910	2,253 („)	6.8	581
1911	3,799 („)	3.4	888
1912	4,076 („)	3.6	857
1913	5,132 („)	2.2	950
	4,917 (9 months, March to December)		

Sources: Dorman and Smith accounts. B. R. Mitchell & P. Deane: *Abstract of British Historical Statistics*, pp. 64–5, 221–2, 367–8, 417–9.

The Northern Society

With the move to Salford, following on their previous technical successes, the status of Dorman and Smith among their professional colleagues in the north-west inevitably grew. It was only then to be expected that both the partners (particularly the extrovert Dorman) would be prominent members of the newly established Northern Society of Electrical Engineers which was formed at the end of 1893. The society was established as an autonomous organisation with the specific purpose of providing a meeting place in the north-west for discussion not only among professional electrical engineers but also with members of the university. The appropriate sentiment was voiced by Dorman at one of their meetings:

> If the professors and the footsole men, as they (are) often familiarly called, could oftener meet and discuss on the same platform, an immense amount of good would immediately accrue to their common market, the world at large.[10]

Dorman became a fairly frequent contributor to the discussions held at the Northern Society[11] with his comments touched by a characteristic sense of humour. For example, the 1895 *Proceedings*

[10] *Proceedings of the Northern Society*, Vol. 2, 1895.

[11] The *Proceedings* only record one contribution by R. A. Smith (on the subject of power transmission), Vol. 4, 1897, pp. 142–3.

PLATE 8. Charles Mark Dorman

PLATE 9. Reginald Arthur Smith

PLATE 10. Herbert George Baggs

PLATE 11. Thomas Atherton

PLATE 12. Capstan section in No 3 shop of the Ordsal Works during the second world war, showing capstan lathes made by Dorman and Smith during the war time machine tool shortage

record his anecdote that 'only last week his firm received a specification of a switchboard and they were amazed to find that its insulation resistance must be infinite. He was rather proud of the estimate of their abilities'.[12]

By 1895 Dorman was already a member of the council of the Northern Society and for the following year, 1896, he was elected joint vice-president with Sebastian de Ferranti. In March 1896 he gave an illustrated lecture to a Society meeting entitled 'The Discharge of Electricity through Rarefied Gases', which was prompted by the recent discovery of X-ray photography. The *Proceedings* record 'the lecture was fully illustrated . . . (with) photographic slides, most of them taken in Mr Dorman's laboratory, illustrating the use of the new photography in surgery, in the distinction of sham jewels, in the detection of the contents of suspicious packages and in other directions.'[13]

The choice of subject might appear unusual – but in fact it was closely related to his main hobby, photography. In turn this hobby had already been put to business account. In the trade press[14] he was referred to as 'one of the most expert amateur photographers in the electrical profession' who applied photography to his business by the illustration of his products. It is almost certainly the case, therefore, that the reproductions, from Dorman and Smith catalogues, which appear in this book, are based upon photographs originally taken by him.

At the end of 1896 Dorman was elected president of the society for the following year. In January 1897, for his inaugural address, he chose to extend the theme of his previous talk by lecturing on the subject 'Radiation'. The object, in his words, was 'to enable the practical man to see that it is not absolutely necessary to use high mathematical analysis in order to obtain a tolerably clear insight into what is generally looked upon as a somewhat abstruse branch of physics'.[15]

It is interesting to record that the president in the succeeding year was J. S. Raworth. It was during his term of office that the

[12] *Proceedings*, Vol. 2, 1895.
[13] *Proceedings*, Vol. 3, pp. 168–77.
[14] *Lightning*, Vol. 7, 18 April 1895, p. 250.
[15] *Proceedings*, Vol. 4, pp. 1–35.

preliminary negotiations took place for the amalgamation of the Northern Society with the Institution of Electrical Engineers. This took place in 1900 and both Dorman and Raworth, in their capacity as past presidents, became members of the first committee of the 'Manchester Section'.[16]

Trade recovery and coming of age

When Dorman was assuming the vice-presidency of the Northern Society the worst of the general trade difficulties in electrical engineering were over and something approaching boom conditions were to be achieved as the turn of the century approached. The principal cause of this change in trade conditions was a sharp increase in public utility investment in electrical equipment – in electricity supply, and in electric tramways and railways. In 1896 annual investment in power stations exceeded £2m for the first time. Yet in the first five years of the twentieth century £33m was invested in public utility supply and by 1903 only two towns in the country with a population exceeding one hundred thousand were without supply. The replacement of horse-drawn trams by electric tramways was largely accomplished between 1897 and 1906. The construction of the London underground system was virtually completed between 1898 and 1907. With the exception, therefore, of steam railway electrification the main period of public utility investment in this field occurred in the decade 1896–1906/7.[17] The evidence of expansion in this sector, however, is complicated by the fact that during the early years of the twentieth century there is some evidence of falling real incomes and rising unemployment in the country as a whole.

Nevertheless, a comparison of the financial results of Dorman and Smith clearly indicates that considerable expansion took place between the recession years of 1893–4 and 1906–7. Between these two dates the wages bill had increased threefold whilst net profits had increased by nearly the same proportion. On the other hand, the more significant comparison may be with the previous

[16] From 1918 it became the North-Western Centre.

[17] I. C. R. Byatt, 'Electrical Products', in D. H. Aldcroft (Ed.): *The Development of British Industry and Foreign Competition, 1875–1914.*

peak of 1889–90. Since then the wages bill had increased sixfold but net profits had only risen by fifty per cent. The profit performance of those very early years had still not been equalled.

During this decade there is some evidence not only of an increase in the level of the firm's output but also of a change in its composition and sales. This reflected both the changes taking place in the electrical industry in general and the direction in which the firm's technical innovations were taking place. Switches, switchboards and cutouts are given even greater prominence in the catalogues from 1901 onwards, while lampholders and related accessories are relegated in importance. Full page illustrations of traction and other switchboards indicate where the main marketing effort was being guided. Further evidence is provided by Dorman's personal pocket catalogue (1906) which lists the firm's main customers in his own handwriting. By far the largest group consisted of fifty-five local authorities, followed by fourteen privately owned electricity supply companies and ten tramway corporations or companies. Of considerably less importance were a small number of shipbuilding concerns (thus retaining their traditional interest) and other electrical manufacturers or suppliers.

The expansion in the level of output from 1895–6 onwards meant that extensions to the original building occupied in 1892 became necessary. Work appears to have commenced in 1895–6 and continued, with intervals, until 1903. Since this followed immediately upon the poor financial results of 1893 and 1894, the liquidity position was initially stretched. However, Dorman relieved the situation by investing in the firm nearly £3,000 of his personal capital from outside the partnership.

By the time of the partnership's coming of age in 1907, the main building had been considerably extended and its outside appearance greatly enhanced (*Plate* 7). The works consisted of two machine shops, two switchboard building shops, a grinding and polishing shop, a fittings assembly shop, a pattern shop, a large glass and china stockroom, a metal and switch stockroom and a brass foundry. The offices, including storeroom, were on the first floor with the switchboard drawing office above. Unquestionably, here was the clearest manifestation of the progress achieved in the twenty-one years since the two young engineers took over their small business in Brazennose Street.

Partnership in transition

Further changes, however, were in the air and these were first evident in the partners' marketing arrangements. In 1907 S. T. Pemberton was appointed agent for the Birmingham area and F. T. Hanks for the Newcastle district. In the following year the London offices were moved from Charing Cross Road to larger premises in Victoria Street. These were placed under the charge of Messrs Gambridge and Ward, who were given responsibility for Greater London and the south coast.

In the early years of the century, Oswald Haes of Haes and Eggars in Sydney was their Australian agent. However, he died suddenly in 1909 when only forty-four years of age. Haes was one of the pioneers in electrical engineering in Australia. He originally came to Australia in 1891 as the representative of the Brush Company. He was a founder member of the New South Wales Electrical Association and was its president on a number of occasions. He gave a number of public lectures on electrical matters both in Australia and New Zealand. He was instrumental in the transfer of electrical supply interests to the Melbourne City Council and was also closely associated with the installation of the electric light system in Sydney.[18] His early death must have been a considerable loss to Dorman and Smith, who had built up a considerable Australian business and who were later to experience difficulties in this quarter.

The period 1908 until early 1910 was one of recession in the electrical industry. In part this was a reflection of a general slackness in trade but it was also due to the termination of a decade of fairly intense activity in the electrical trade. Dorman and Smith's sales did not decline but their profit margins were sharply reduced, particularly in 1908–9, resulting in a considerable reduction in net profits. By 1910, however, the trade was well on the way to recovery. A shipbuilding boom was under way

[18] *Electrical Review*, Vol. 65, 1909, October 22, p. 665 and October 29, p. 705.

and unemployment in the engineering trades fell to a very low level. The period 1910–14 was, therefore, the most prosperous yet in the history of Dorman and Smith. The volume of sales increased sharply, profit margins widened and, as a consequence, net profits in 1913 were over three times greater than they had been in 1909.

The onset of a further period of expansion necessitated further extensions to the working premises which occurred during the year 1911–12. But it also raised a more fundamental issue – the future of the partnership itself. As the works were extended, labour force increased and output expanded, the volume of work falling on senior management became progressively greater. The partners, however, were no longer young men. By 1914, Smith was fifty-seven years old, Dorman four years younger. The problem of management succession, upon which many partnerships have foundered, would soon raise its head. Of the two partners, only Dorman had a son, but he was only in his early teens. It was in these circumstances that H. G. Baggs (*Plate 10*), the works manager, came to occupy a more central role in the business affairs of Dorman and Smith. To pursue this matter further, it is necessary to examine the initial terms of the partnership and the ways in which they were subsequently modified.

When the partnership commenced on Lady Day 1886, Dorman paid £1,000 into the partnership and Smith paid £500. The provisions relating to the allocation of net profits between the two partners were in three parts:

1. Each partner was to receive five per cent annual interest payment on the capital he had invested in the business,
2. After deduction of interest payments each partner was to receive one-third of the remaining net profits,
3. The remainder of the net profits was to be divided between the partners in proportion to the capital each had invested in the business.

The first element was a straightforward interest payment, the second corresponded to a management fee linked to the net profit of the business and the final element was equivalent to a dividend on capital invested.

The capital invested in the business by a partner (upon which two elements in the division of profits were based) consisted of the original and any subsequent sums he paid into the partnership *plus* his accumulated share of the net profit *less* his drawings on the partners' account for personal expenses. The provisions therefore provided for the initial ploughing back into the business of the entire profit earned and encouraged personal frugality if a partner wished to maintain his proportionate share of the profits.

Whilst Dorman initially placed a greater capital sum in the partnership, he drew upon his personal account to a greater extent in the early years (though, since profits were very low at this stage, it could hardly be taken as a sign of extravagance); Smith also made two small new investments in the business with the net result that by 1889 he had a greater financial interest in the business than Dorman and received the greater share of net profit. However, for reasons previously recorded, Dorman in 1894 invested nearly £3,000 as additional capital in the business and, at the same time, the terms of the partnership were modified. From this date only the interest payment component was to be related to the amount of capital invested by each partner. Once this amount was deducted from net profits, the remainder was to be divided equally between the two partners.

Finally, in 1907, a further change was made in the arrangements which in part reverted to the original principle of the partnership agreement but had the incidental effect of relating H. G. Baggs's payment to the net profits of the business.[19] First, interest was charged at five per cent on a nominal £7,000 and the proceeds were divided between the two partners in proportion to their average capital holding in the business in each of the twelve preceding months. Then one-fifth of the net profit less interest payment was credited to Baggs as a management fee.[20] Thirdly, after this amount was also deducted, each of the partners received one-fifth of the residual as their management fee. Finally, the

[19] Not only was this the year in which Dorman and Smith came of age, it was also the time when private company legislation was approved by Parliament.

[20] Baggs's management fee was debited to the trading account and the net profit figure in the accounts was reduced by an equivalent amount.

amount of net profit remaining was divided between the two partners in proportion to their capital holding.

From the beginning of 1897 Baggs had received a commission calculated on the total sales of the partnership but these latest provisions were of far greater significance. The sums due to him were credited to a special personal account, forming part of the partnership's set of accounts, and was subject to drawings in the same manner as the personal accounts of the two partners. As the profit record of the business sharply improved in the 1910–14 period, the capital holding of Baggs within the partnership increased commensurately and reached a very substantial level.[21] This capital holding was not eligible for either interest or dividend payments but such an incongruous situation would not exist for long. In 1914 the partnership was changed into a private company enjoying limited liability and Baggs's holding in the partnership became a shareholding in the new company. The impending problem of management succession had apparently been solved and the partners gained the security of limited liability and the advantages of corporate status in the process.

An assessment

At the outbreak of the first world war, Dorman and Smith had many reasons to be satisfied with their achievements over the previous twenty-eight years. They had acquired a new legal status, greatly enlarged and improved work premises and business assets twenty times greater than those they had started with. They were firmly established, with an international reputation for high quality switchgear and ancillary lighting equipment. They had played an important role in the rapid technical progress in their section of the electrical industry, most notably in the introduction of porcelain in switches and lampholders but also in developing enclosed switchboxes and conduit ceiling roses.

On the other hand, their commercial and financial achievements after 1890 were not as outstanding. Whilst the growth rate of the partnership was considerable, the rate of growth in the

[21] Over £2,000.

industry as a whole was substantially greater. For example, local authority receipts from electricity supply increased from £0.1m to £5.4m between 1895 and 1914, yet Dorman and Smith's net assets only increased threefold during the same period. The net profit record was less impressive. From 1890 until 1910 the average annual rate of profit increase was very modest and, even in the exceptional conditions of 1913, profit was only 2½ times greater than twenty-three years earlier when production was still based at Brazennose Street.

Two features of significance stand out in the financial records. First, the rate of return on capital employed, which was exceptionally high in the pre-Salford period, fell very sharply after the move and continued at or below this level for most of the period (see Table 5 below). A rate of return on total net capital employed of approximately ten per cent must be regarded as quite modest, especially as this has been calculated before the deduction of the partners' management fees. Since the partnership relied mainly on ploughed-back profits to finance further expansion, this profit record must have acted as a brake on the growth rate – assuming that profitable avenues for expansion existed and that the partners wished to take advantage of these.

TABLE 5 DORMAN AND SMITH PROFITABILITY, 1888–1913

Year	Net profits as percentage of total net assets	Net profits as percentage of total fixed assets
1888	26	150
1890	28	212
1894	9	28
1904	10	26
1909	6	16
1912	10	36
1913	13	43

Source: Dorman and Smith accounts.

Secondly, the disposition of assets particularly in the later years revealed a heavy emphasis on stock, sundry debtors and cash assets (see Table 6). Between 1904 and 1913, for example, the value of net fixed assets remained virtually unchanged and the thirty per cent increase in total assets which occurred was almost entirely due to increases in the three items mentioned. In part this can be attributed to special factors – for example, the need to maintain a sufficient degree of liquidity to meet the personal drawing requirements of Dorman, Smith and Baggs – but not entirely. In strictly commercial terms it is open to question whether the partnership was making the fullest and most effective use of its capital assets.

TABLE 6 DORMAN AND SMITH ASSET STRUCTURE, 1888–1913

Year	Total net assets (*£ thousands*)	*Of which the main items were:* Buildings, plant, tools, fittings and fixtures	Sundry debtors	Stock	Cash and securities
1888	3.5	0.6	0.9	1.1	0.7
1890	8.4	1.1	4.0	2.7	0.5
1894	13.1	4.7	2.7	4.9	0.6
1904	30.6	11.0	6.4	10.2	2.7
1905	33.4	10.6	4.4	14.0	4.3
1912	39.0	11.4	7.7	13.7	5.9
1913	44.7	11.9	8.9	13.5	10.5

The partners may well have been satisfied so long as they were able to maintain a modest rate of increase in the volume of business and level of profits. Nevertheless, these financial indicators contain a warning of potential difficulties ahead, particularly when general trading conditions deteriorate.

CHAPTER THREE

The Private Company 1914–37

The private company of Dorman and Smith Limited was registered on 29 January 1914, acquiring the assets previously owned by the partnership valued at approximately £36,000.[1] The articles of association provided for the appointment of a minimum of two and maximum of five directors. Each director was required to hold preference or ordinary shares to the minimum nominal value of £1,000. First managing directors (named in the articles of association as C. M. Dorman, R. A. Smith and H. G. Baggs) were specially protected in their security of tenure, so long as they retained the necessary financial holding in the company. In addition, Dorman and Smith enjoyed the status of permanent governing directors, which made them totally immune from the threat of removal so long as they retained their financial interest in the company. Managing directors were required to give their 'whole time and attention' to the business of the company but permanent governing directors, on retirement from full time employment, could continue as governing directors. In such circumstances each governing director would receive one half of the managing director's salary. Each of the three first managing directors was entitled to one-fifth of the annual net profits where these exceeded £2,500 but, in any event, they would each receive a minimum amount of £500, which would be a first charge on the finances of the company.

Initially five directors were appointed – the three first managing directors and the wives of the two permanent governing directors, Cordelia Dorman and Edith Eleanor Smith. The first

[1] The lower valuation of assets in the company than in the partnership at the end of 1913 is assumed to lie in the substantial withdrawal of cash and security assets by the partners immediately prior to the establishment of the private company.

chairman of the company was C. M. Dorman, and R. A. Smith became its deputy chairman. The nominal share stock consisted of 35,000 ordinary shares of £1 denomination and 15,000 preference shares of the same denomination.[2] Of these, 15,000 ordinary shares and 10,000 preference shares were issued as follows:

TABLE 7 DORMAN AND SMITH SHAREHOLDERS, 1914

	Ordinary shares	Preference shares	Total
C. M. Dorman	6,300	—	6,300
R. A. Smith	5,700	—	5,700
H. G. Baggs	3,000	3,000	6,000
C. Dorman	—	4,200	4,200
E. E. Smith	—	3,800	3,800
TOTAL	15,000	11,000	26,000

The war years 1914–18

The company came into being during a boom in the electrical trade and at the most prosperous point in its history to that date. However, the onset of war provided it with its first test of adaptability. Amongst the earliest problems which confronted the company was the disruption of its overseas trade and the delicate ethical and legal situation arising from existing contracts with German-owned undertakings. R. A. Smith, in a letter to *The Electrician*[3] indicated the nature of the dilemma and the manner in which he believed it should be handled:

> ... we further note that under Clause 5 it would be a contravention of the law to supply goods for

[2] In addition, each of the original partners made a short term loan of £2,500 to the company at its inception.

[3] *The Electrician*, Vol. 73, 18 September 1914, p. 943.

> the use or benefit of an enemy. We have before us at the moment the status of A.E.G. Electrical Co. (Ltd) and as we are informed that the bulk of these shares are, or were quite recently, held in Germany, we have decided to discontinue any business relationship with this company in compliance with the above clause. We are, however, threatened with penalties by A.E.G. Electrical Co. for the non-execution of an order for which we have tendered, and we are asking for the support of our competitors in the attitude we have taken up.

Apart from immediate problems of this kind, the major task of the company was to redirect its production facilities towards the war time effort. In view of the long association with the provision of electrical equipment for ships, it was not surprising that the greater part of their work came to be undertaken for the Admiralty. In certain cases this meant, as the Patent Records testify, applying their inventive skills in new directions. In 1915 a patent was taken out for depth indicators and, in 1919, for a device used in cutting and sweeping the cables of marine mines.

In addition the company had to adjust to the now familiar situation of rapidly rising material and labour costs. Between 1914 and 1919 the cost of materials purchased increased by over sixty per cent whilst the wages bill increased by nearly 150 per cent and both these increases were more due to increased price levels than increased activity. This presented two related kinds of problem.

In the first place, the working capital required by the company increased, thus placing pressure on its cash reserves. The problem was further aggravated by the change in financial arrangements following the adoption of company status. The partners had previously retained their share of the profits in the business, only drawing upon it when necessary for their personal requirements. With the establishment of the company, the profits were paid out as dividend and interest (to the extent that they were distributed) and so eliminated one source of working capital. Faced by increased cash demands, Dorman and Smith increased their combined loans to the company from £5,000 to £6,000 during 1915–16. In 1916 the three managing directors loaned £2,550 accruing to them as dividends. Then in 1917, it was decided to convert these loans into a longer term source of finance by the issue of additional shares of which Dorman took 1,200, Smith 1,100 and Baggs 700. Finally, in 1918 Dorman purchased a further 800, Smith 600 and Baggs 600 ordinary shares. In this way the company was able to finance its increased working capital requirements without resort to bank overdraft.

At the same time the company was confronted with the related problem of adjusting its price lists sufficiently quickly in the face of rapidly rising costs of production. By October 1918 virtually all prices had been increased by over 100 per cent of their pre-war levels and fuse wire had been increased by 500 per cent. At times price increases were made with bewildering frequency. For example, prices of unmounted handle fuses were increased to 85 per cent above the pre-war level on 1 April 1918, to 95 per cent on 1 June, to 115 per cent on 1 September, and to 125 per cent on 1 October 1918.

However, earlier in 1915 and 1917 the unavoidable delay in adjusting price lists resulted in smaller profit margins and was the major reason for the poorer profit performance in those years.

Nevertheless, taken as a whole, the profit performance of the company during the first world war period was quite satisfactory without, in any sense, being outstanding. Because of the change in the treatment of management payments it is not possible to make a strict comparison between pre-1914 and post-1914 profit data. However, the absolute level of profits during the war, with the exception of the two years referred to, was of the same order as in the boom conditions immediately prior to 1914. Throughout this period the customary 5 per cent was paid on the preference stock and 12½ per cent or 15 per cent was paid on the ordinary stock in addition to the profit-related management expenses which were paid to the three managing directors. Yet though the value of sales increased considerably during these years, the profit margin was much lower than immediately prior to the war. No dramatic change in the fortunes or scale of operations of the company occurred at this time.

Post-war boom and management succession

The war time period witnessed an eighty per cent increase in the consumption of electricity, but because of controls there was only a modest injection of new capital into the industry. With the restoration of peace time conditions it was to be expected that the frustrated demand for electrical equipment would be released, creating boom conditions in the industry. This is precisely what occurred during the first two years of peace; a substantial increase in the level of activity, a sharp increase in dividend levels and a considerable injection of new capital into the industry attracted by the high returns. Dorman and Smith Limited participated fully in this expansionist phase. In 1920–21 the value of sales was forty per cent higher than in 1918–19 and net profits were nearly three times higher than in any previous year.[4]

However, behind these 'heady' trading conditions, changes of greater long term significance were occurring in the company. Shortly following the restoration of peace, at a meeting of the board of directors on 20 November 1919, R. A. Smith announced his intention to retire from the position of managing director and to continue only as a governing director. Earlier in the year, no doubt in anticipation of this change and the increased burden expected to fall on the two remaining managing directors, W. T. Tallent Bateman was appointed works manager. Probably the adjustment alone in the management hierarchy could have been made without too great a difficulty to the company. However, on 12 February 1920, C. M. Dorman died following a heart attack whilst on holiday at Llandudno. Within the space of little more than two months the firm had lost the full time services of its two founders. In fact, C. M. Dorman had been in ill health for some time. Mention was made in the previous chapter of his son, Harold Mark Dorman, known affectionately to his parents as 'Micky', who was then studying as an engineer at the University of Manchester – no doubt with the ultimate intention of assuming a position in the company. In

[4] A substantial part of the increased profits had to be paid in excess profit levy, the remainder being placed to reserves.

1916 H. M. Dorman joined the Royal Engineers and in September 1917 was stationed at Marlow prior to embarkation for France. On a Sunday evening walk, when returning to barracks, he missed his way, fell into the river and was drowned. He was formally identified by H. G. Baggs who recognised a drawing compass and wire gauge in Dorman's possession as former presents to the young man. The death of his son, then aged nineteen years, in such tragic circumstances, greatly affected Charles Mark Dorman. For some time after this event he was away from work and apparently never regained the same zest for life as he had previously possessed. His premature death was regarded as partly due to this sad event.[5]

As was to be expected, R. A. Smith assumed the chairmanship of the company and for a two month period was confronted by quite complex financial arrangements arising from the death of the major stockholder and made more difficult by the exceptionally high working capital requirements during the peak of the electrical equipment boom. Earlier, in February 1920, Smith loaned £500, Baggs £500 and Bateman £1,000 to assist the short term finances of the company. On the death of Dorman, Bateman was invited to become a director of the company and his £1,000 loan was converted into an ordinary shareholding.[6] In March 1920 and again in September 1920 the dividends accruing were loaned back to the company by the shareholders because of their stated reluctance to finance the increased working capital requirements by bank overdraft.

By November 1920 the loans outstanding from the estate of C. M. Dorman amounted to £4,000. At the request of his widow (probably in accordance with the terms of the will) this amount was equally divided between five relations. In turn the relatives were invited to continue the loans to the company and the interest rate was raised to 8 per cent as a further inducement. Then in March 1921 Mrs Dorman indicated her desire to sell

[5] *Electrical Review*, 20 February 1920, Vol. 86, p. 244.

[6] Bateman's management fee was calculated as one-tenth of net profits. Since R. A. Smith, on retirement, was only eligible for half of his previous fee percentage, net management fees as a proportion of net profits fell by one-third as compared with the immediate post-1914 period.

PLATE 13. The Ordsal Works in 1958 at the time of the move to Preston. The buildings in the foreground and centre are additions to the original factory

PLATE 14. Dorman and Smith directors with guests, 1951. *Left to right*: Richard Amberton, Alderman E. A. Hardy JP, MP, B. L. Cooper, Alderman F. Cowin JP, C. M. Nesbitt, T. Atherton, M. L. Cooper

£2,000 of her husband's ordinary shares. These were independently valued at 108¼ per cent ex dividend and were purchased by R. A. Smith (1,000), H. G. Baggs (600) and Tallent Bateman (400). It is interesting to note that seven years after the formation of the company, at a period of peak activity in a major growth industry and where general price levels had increased greatly, the share value was only 8¼ per cent above its original nominal value.

H. G. Baggs

After sorting out the financial intricacies resulting from his former partner's death, R. A. Smith resumed the role of governing director, resident in the Lake District, paying relatively infrequent visits to the works and only attending the more important board meetings; from this point in time until 1937 Herbert George Baggs is both first managing director and, *de facto*, chairman of the company. As such he exercised a dominating influence over its fortunes for the next sixteen years. What sort of a man was he?

He joined Dorman and Smith at the beginning of 1894 at the age of twenty-four years. He was appointed as works manager and became a joint managing director in 1914. It is assumed that previously he was apprenticed in an engineering works but no details of this are available. A short while after joining the partnership he became a member of the Northern Society of Electrical Engineers. He transferred from that society in 1900 to become an associate member of the Institution of Electrical Engineers, later becoming a full corporate member. In his early years with the partnership he was associated with a number of technical developments and his name is recorded, on a number of occasions, in the Patent Records.

By those who knew him he was acknowledged as a very good engineer and draughtsman who continued the tradition of manufacturing high quality products previously established by Dorman and Smith. In the words of a later chairman of the company 'he never made a shoddy product in his life'.[7] Whilst he did not possess the innovative qualities of his predecessors, his technical abilities were certainly equal to the requirements of a medium sized electrical engineering undertaking. However, additional qualities are required in a senior managing director and it is in these respects that signs of weakness were evident.

In the first place there was a clear tendency, under H. G. Baggs's direction, to emphasise the importance of quality in the firm's products without sufficient regard to what was commercially most in demand. The qualities of the 'commercial engineer', discussed in the previous chapter, were not sufficiently in evidence and this inevitably affected the sales performance of the company. The company became technically more conservative whilst maintaining its previous tradition of financial conservatism. Secondly, Baggs could become irritated and short tempered with his staff if they failed to match up to the high standards he expected of them. In an abrasive atmosphere, changes at senior management level were too frequent for the healthy development of the company.[8] When the general economic climate deteriorated at the end of the 1920s, the company was ill equipped to weather the storm.

Yet, according to one of his former employees, away from the pressure of work, he could be friendly and gentle, with a keen and cultivated interest in painting and yachting. At one time he was a prominent member of the Royal Mersey Yacht Club and of the West Lancashire Yacht Club in which he held the post of official measurer for some time. He was a yacht designer of some reputation; in 1898 he collaborated with Mr W. Scott Hayward to design the *Seabird* class of yachts, of twenty feet overall length. The class is still being built and raced today. Baggs also designed wireless sets. Clearly he was a man of many talents which, regrettably, did not always extend to the effective management of men or the exercise of entrepreneurial skills.

[7] Thomas Atherton: Address on the sixty-fifth anniversary of the founding of Dorman and Smith.

[8] Certain of these changes may have been highly desirable. It is the number of changes, in relation to the size of the senior management group, which is emphasised.

Management changes 1921–37

The weaknesses of the company both in technical development and marketing were recognised as early as 1922 when the post-war boom had just ended. In April of that year the board of directors recorded that the volume of work during the previous two years had left insufficient time for the improvement of the company's products and selling organisation. It was, therefore, decided that Tallent Bateman should relinquish his duties as works manager and devote all his time to the development of new products. This, however, does not appear to have been a successful move. In July 1923 Bateman intimated his intention to resign from the company, though the form of his letter of resignation would indicate that he was encouraged to take this course of action. In February 1924 it was decided to appoint the secretary to the company, J. A. Wilkins, as a director. Wilkins had been the secretary since the company's inception and was highly regarded in that position. However, he was not equipped to rectify the technical deficiencies which had been noted in 1922.

As far as can be established no action was taken on this matter until November 1926 when the board resolved to offer the post of 'Electrical Engineer of the Company' to J. H. Beaumont Robbins. The duties were listed as being 'responsible in the way of developing new work and in dealing with the company's enquiries and productions. The appointment would be nominally that you would act as the Electrical Engineer of the Company and that in effect you would be the chief assistant to the Managing Director' (*ie* H. G. Baggs). The terms of the appointment implied that there would be a 'probationary period' until March 1928 when, if satisfaction had been given, the rate of remuneration would be significantly increased. However, in July 1928, J. H. A. Spaink was appointed electrical engineer to the company on similar probationary terms to those offered to Beaumont Robbins, who apparently had left the company.

On 16 October 1930 R. A. Smith died, the interment taking place at Stoney Middleton, his place of birth. In consequence his shares passed to his widow and H. G. Baggs became the chairman of the company. In April of the following year he was faced with the possibility of losing his works manager, S. H. Chatterley,[9] who had been offered a position in Glasgow. It was, therefore, decided to offer Chatterley a directorship. In August 1932 the board decided to terminate the appointment of J. A. Wilkins and this is known to have arisen from a personal disagreement between Baggs and Wilkins. With this dismissal Mr Thomas Atherton, previously assistant secretary, was appointed secretary to the company.

By this time the country was in the depths of economic depression and sales were at the lowest level in the lifetime of the company. In December 1932 Sambidge, the London agent, forwarded his resignation as from June 1933. The new London agent was Preston and Company but apparently this new arrangement was not particularly satisfactory. In December 1934 it was decided to terminate the agency and Preston was replaced by Amberton and Partners.

In October 1934 it had been decided that because of the increase in the wages bill, disproportionate to the increase in turnover, 'the wage list should be revised and at least six of the older workmen be replaced by a larger number of young boys and girls, so restoring the proportion to a more economical rate'. This apparently did not prove as economical as supposed. At the board meeting in May 1935 it was recorded that 'we have lost some of the older and more experienced members of the staff during the last eighteen months and it has consequently been more difficult to carry on with the same efficiency and economy.'

In November 1935 a Mr Capp was appointed as works manager and S. H. Chatterley became sales manager. Thomas Atherton was appointed to the board in March 1936 – the second secretary to serve in that capacity. Meantime, Chatterley's occupancy of the sales manager's position was short lived. In May 1936 he wrote 'expressing his willingness to withdraw from the board of the Company to make way for another Director'. In July 1936 Major Amberton, of Amberton and Partners, was appointed a director of the company.

[9] It is not clear whether Chatterley had previously replaced Spaink or was appointed separately. No further reference to Spaink can be found in the records of the company.

This catalogue of management changes underlines the weaknesses to which the company was prone during this period, especially in two major areas of operation – the development of new products and its selling arrangements. It is now intended to examine the effect that these weaknesses and the general economic climate had on the financial results of the company during this sixteen year period.[10]

Financial results during the 1920s

Dorman and Smith's profit performance passed through a very pronounced cycle during this decade. In 1921–2 profit levels were already falling sharply from the very high level of the previous year. In the following two years, 1922–4, profits were at the lowest level since 1914. There was a significant recovery during 1924–5 and a further very substantial improvement during 1925–6 to a peak of £8,700; a level only previously exceeded in the boom year 1920–21. The profit level fell slightly during the following year (though still relatively high); thereafter profits fell continuously in each successive year, reaching only £3,000 in 1929–30.

During the first half of the 1920s the cyclical movement in the company's profits mirrored the pattern of fluctuations in industry generally. From 1921 to 1924 there was a continuing decline in the ordinary share index and in the average rate of dividends paid by electrical manufacturing companies. Then both indicators substantially improved during 1924–6 only to fall slightly in the following year. Thereafter the pattern of fluctuation diverged. In industry as a whole, as well as in the electrical manufacturing sector, continuous expansion occurred, reaching a new peak at the end of the decade. This is in marked contrast to the progressive decline in the profits of the company during these same years.

The difficulties experienced by Dorman and Smith Limited in the late 1920s were partly due to problems in overseas markets, especially a fall in fuse sales to Australia and New Zealand.

[10] See Table 9 at the end of this chapter for the statistical details of the company's financial performance.

However, there is reason to believe that these difficulties were more deep seated and long term in nature. Between 1920 and 1930 sales of electricity increased nearly threefold whilst local authority expenditure on electricity supply doubled. During the same period the issued capital of electrical manufacturing companies increased by fifty per cent. Yet in the case of the company, the value of its sales in the late 1920s was no higher than a decade earlier and the same was true of the level of its profits. Though within a rapidly expanding industry, the company was stagnating. Judged against this background, the contraction in its profits at the end of the decade, when industrial activity was generally expanding, was particularly ominous.

A financial characteristic of the company in this period, as at earlier times, was its cautious attitude towards cash reserves. In part this was justified by the high level and degree of variability in the debts of the company's customers. Inevitably, however, this had a depressive influence on the rate of return on capital employed by the company. On the other hand, despite the longer term stagnation in the level of its sales, Dorman and Smith Limited did expand the land and building assets at its disposal. In December 1926 the board of directors decided, because of the problem of storing empty packing cases, to enter into negotiations for the purchase of a plot of land, bounded by Hampson Street and Middlewood Street from the LMS Railway, as an extension to the Hampson Street Works acquired in 1915. Later, in September 1929, the company made a further purchase of land, housing and stabling adjoining the Hampson Works from G. W. Andrews, coal merchant. Yet in contrast, during the same period there were no significant additions made to machinery, fixtures and fittings, and the written down value of these items in the balance sheets of the company continuously fell. This was yet another indication of the underlying weakness in the company's position.

The great depression and its aftermath, 1930–37

The end of 1929 saw the onset of a sharp and unprecedented recession in the British economy which reached its lowest point in 1932 and showed no significant recovery until 1934. This depression adversely affected the electrical manufacturing and electrical supply industries, especially in those parts of the country where it had the most serious impact. Nevertheless during this period the electrical industry continued to develop and during the latter half of the 1930s resumed its previous rapid expansion. Between 1929–30 and 1935–6 electricity sales to consumers increased by more than eighty per cent and in no year, even at the lowest point in the depression, did electricity sales decline. The average rate of return on issued capital by electrical manufacturing companies fell from its peak in 1929–30 to its lowest level in 1933–4 but by 1936–7 it had more than regained its original level.

With the dramatic reversal in general trading conditions, the underlying weaknesses in the company became apparent. Sales fell from £70,000 in 1929–30 to only £30,000 in 1932–3. Because a corresponding reduction in expenditure could not be achieved, losses were incurred in each year from 1931 to 1935. Though the reserve fund was exhausted, even so there was insufficient for payment of preference stock interest, which fell into arrears from April 1931 onwards. Between 1929 and 1936 no additions were made to land, buildings, machinery, fixtures or fittings. Stocks fell from £22,000 to £9,000, cash assets from £9,000 to £2,500 and total assets by over one quarter.

The depression financially affected the company in both a direct and indirect way. As the chairman remarked at the annual general meeting in 1931, 'scarcity of orders and extreme competition frequently necessitated work being booked, solely with the object of keeping the works employed, at prices quite unremunerative'. But, in addition, as the effects of the depression became transmitted internationally, tariff barriers were erected. As a consequence, the company suffered a sharp reduction in overseas sales – between 1930 and 1932 Australian trade fell from 16 per cent to 3 per cent of its total sales.

The general industrial revival after 1933 led to some improvement in the company's position but initially progress was very slow. Small losses were made between 1933 and 1935 and this was converted into a very modest surplus in 1935–6. During

1936–7, as the peak year in the 1930s approached, the profit performance improved substantially to a level commensurate with the later 1920s. Yet this was little different to the profit levels immediately at the end of the first world war when management fees were very much higher. Indeed, the value of sales and of net assets was only slightly greater than at the time when the company was established, twenty-three years previously. Whilst the industry had grown to maturity during this period, Dorman and Smith Limited was unaltered in scale and considerably weakened in competitive position.

Thomas Atherton and the reconstruction of the company

Whilst the company had weathered the storm it was still in a very weak position. The underlying deficiencies in the company during the 1920s, far from being corrected, had become more plainly evident and had hardened with the passage of time. With the next reversal in general business activity the company was in real danger of plunging into deficit again. In the early 1930s the effects of the depression had been partially cushioned by generous cash reserves. These no longer existed to provide significant assistance for a second time.

By the beginning of 1937 H. G. Baggs was sixty-seven years old and had no son to follow him in the business. He did not possess the temperament or organisational powers required to place the company on a long term viable basis. With the resumption of a modest profit in 1936 this was the time for Baggs to extricate himself from the company before another downturn in general economic activity occurred. Where, however, was an acceptable purchaser to be found?

In November 1933, at the height of its financial difficulties, the company had received a communication on behalf of a client wishing to purchase a controlling interest but negotiations did not proceed. At the beginning of 1937 the company received a further inquiry, of a preliminary nature, from a financial holding company in London. As events were to turn out, however, the offer to purchase came from one of the newly appointed directors of Dorman and Smith Limited – Thomas Atherton, the company secretary.

Thomas Atherton (*Plate 11*) was born in St Helens on 1 July 1899. He completed his school education in 1914 at Hindley and Abram Grammar School. He then took a position with John Preston, a Wigan solicitor, with a view to becoming articled. A year later he joined the Hindley Field Colliery Company Limited as a commercial trainee and in the next ten years served in the buying department and the general office. In 1917 he joined the Navy for the remaining duration of the first world war, returning to the Hindley Field Colliery Company Limited in 1919. In 1924 he was successful in obtaining associate membership of the Chartered Institute of Secretaries and associate membership of the London Association of Accountants. Following an interview with H. G. Baggs he was appointed to the position of cashier in Dorman and Smith Limited from 25 March 1925.

At a meeting of the board of directors in September 1926 it was decided to appoint Thomas Atherton as assistant secretary to the company so that he could formally deputise for J. A. Wilkins in the latter's absence at board meetings. Then (as previously described), following a dispute J. A. Wilkins was dismissed in August 1932 both as director and company secretary and Thomas Atherton was appointed company secretary in his place. This occurred at the most critical financial point in the company's history. Immediately monthly meetings of the board were initiated (previously they had occurred less frequently) at which the new secretary presented monthly trading statements of the company's business. The financial year 1935–6 was the first year in which the company had made any profit for a number of years and at the end of that year Thomas Atherton was invited to become a director of the company. His shareholding (1,000 ordinary shares) was obtained without payment by transfer from H. G. Baggs on agreement that he held these shares as a trustee and would retransfer them on demand and pay all dividends to H. G. Baggs.

With twelve years practical experience with the company behind him, Thomas Atherton was well aware of its weaknesses. But he also had an appreciation of its latent potential as well as a personal feeling for the name of Dorman and Smith and for the high quality work with which it was traditionally associated.

In a letter to his financial advisers in 1937 he summarised his assessment of the company in the following terms:

> (The Company) has a wonderful name and reputation internationally, for the production of sound products[11] but is considered justly, to be out of date. There is no reason why sales should not be very considerably increased if our methods and manufactures are brought more into line with modern ideas and requirements.

Thomas Atherton was not an engineer, and to this extent an important break with tradition was to be made. But in appointing Thomas Smalley he was soon to gain a very capable engineer. What was then required was a redirection of policy towards products that maintained the Dorman and Smith tradition of high quality but more closely met the market's requirements than in the past, and an improvement in work morale through the effort, enthusiasm and encouragement of the managing director. This, based upon a shrewd appreciation of the commercial and financial situation of the company, was to be Thomas Atherton's main contribution and it was to open a new chapter in the development of the company.

Inevitably time would be required to reorganise the company and place it in a new position of strength; the premature onset of recession could have been a deathblow to the company. The first calculated gamble by Thomas Atherton was that the prevailing buoyant trading conditions would continue sufficiently into the future. In this he showed an early appreciation that even if the domestic boom broke, rearmament was under way and would provide considerable support for the order book from the Admiralty. Writing to his bankers at a critical stage in the negotiations in February 1937 he noted 'much as we both, as taxpayers,

[11] The established quality of Dorman and Smith's products meant that they were frequently used for 'prestige' and special quality work. Recent contracts, at that time, included work for Buckingham Palace, St James's Palace, Royal Courts of Justice, Bank of England, Royal Bank of Scotland, Earl's Court Exhibition, Bodlean Library (Oxford), King's Hall (Belfast), the *Queen Mary*, Malta Dockyard and Singapore Naval Dockyard.

deplore the necessity of the recent Government decision to borrow £400m for the provision of armaments, that decision makes absolutely certain at least five exceptionally prosperous years for the company.'

The second calculated gamble was of a purely financial kind. At the time when he opened negotiations to acquire the company, Thomas Atherton possessed only £100 in cash assets and held his director's shares in the company merely as trustee for H. G. Baggs. The issued capital of the company at that time was £34,000 and the net book value of its assets was £38,000. He therefore had to raise a very substantial sum of finance and engage in a reconstruction of the company requiring the support of a number of potentially difficult interested parties. The details of these negotiations are of considerable significance in the company's historical development; in addition they provide insight into the attributes of the man who was to assume control of the company.

The financial strategy adopted by Thomas Atherton was based initially upon the experience of an acquaintance of his who had acquired control of a company using the options approach. At the beginning of January 1937 he first approached H. G. Baggs, inviting him to sell the shares he held and those of his nominees. This covered 6,800 ordinary shares and apparently included the holdings of H. G. Baggs himself, his daughter B. M. Baggs, Amberton, Aikmen and two thousand shares held by Bleakley and Bleakley, executors of Mrs E. E. Smith's will, on behalf of Llewllyn Urban Smith, one of the beneficiaries.

In a letter dated 15 January Thomas Atherton made a provisional offer to H. G. Baggs, conditional upon Dorman's widow being prepared to sell her holding. H. G. Baggs and his nominees were to be offered £2 for each ordinary share and £1 for each preference share whilst Mrs Dorman was to be offered £1 for each ordinary and preference share held. In the case of the Baggs holding there was an additional provision for financial benefit from any capital gain should the company be resold within the following ten years. By a letter of the same date, H. G. Baggs indicated his agreement in principle to the terms of the offer and also the approval of Mrs Dorman.

TABLE 8 DORMAN AND SMITH SHAREHOLDERS, 1937

	Preference stock	Ordinary stock
H. G. Baggs	2,000	3,200
B. M. Baggs	—	500
R. Amberton	5	1,000
A. Aikmen	—	100
Cordelia Dorman	4,200	6,200
J. A. Wilkins	—	1,000
M. L. Bleakley & J. F. Bleakley (Executors of the will of Mrs E. E. Smith deceased)	3,800	10,000

Armed with this provisional agreement to his acquisition of the majority shareholding in the company, Thomas Atherton then used the £100 at his disposal to secure the services of Messrs Hall & Company, solicitors, and arranged an appointment at the head office of the District Bank on 20 January. The District Bank had been bankers to the firm of Dorman and Smith for over fifty years and the purpose of the meeting was to obtain financial assistance against the security of the company's assets, in a manner which would enable Thomas Atherton to purchase the above mentioned shares. The scheme proposed was ingenious. First of all it was contingent upon the directors being empowered, without reference to a general meeting, to issue debentures to the amount of the capital required. These debentures would be issued to the bank and the money would be lent to the purchaser on the security of the shares he had acquired. The purchaser would then pay to the company the debenture interest charged by the bank. The amount asked for was £26,000 which would be repaid at a minimum annual rate of £2,000.

The bank, however, did not like the scheme as it stood. First, they considered that the more appropriate method was to form a new company which would purchase the assets of the old company. Secondly, they were concerned that this would become a 'one-man business' and, in the event of Thomas Atherton's death, this would place the bank in a difficult position. In a letter dated 15 February Thomas Atherton pursued this line of inquiry further and asked whether the passive partner that the District Bank had informally mentioned was an electrical engineer and whether he would wish to take any active interest in the company.

This was shortly followed by a second meeting at the District Bank on 19 February. The main lines of agreement were reached at this meeting. A new company was to be formed and the bank would loan £22,500 (later raised to £27,500) at 4½ per cent interest. Repayments of £2,500 at six-monthly intervals would be made until the overdraft was reduced to £15,000. Thomas Atherton was required to take out a life insurance policy for £10,000 for ten years (to cover the bank in the event of his death) and it was recommended that J. Noel Haworth FCA (Thomas Atherton's accountant) be a member of the new company's board. Apparently the bank did not insist that the new investor should own ordinary shares or be a member of the board.

Negotiations had now to proceed on two fronts – in purchasing the holding of those shareholders in the existing company with whom agreement had not yet been reached and with the proposed additional investor in the new company. On 24 February H. G. Baggs wrote to J. A. Wilkins informing him of his own intentions and advising him to make an offer to sell to the majority shareholder. A meeting with Thomas Atherton's accountants was arranged for 8 March when it was hoped that Wilkins would agree to sell at a price of approximately ten shillings per share. Apparently the meeting did not go as planned for, on 21 April, Thomas Atherton wrote to J. A. Wilkins suggesting a personal meeting and eventually a price was agreed at only slightly less than £1 per share. An approach was made for the Bleakley shares on 2 March and the executors agreed to sell at 17s 6d per share but this was not acceptable. Later in March the whole Bleakley shareholding was sold for £9,315.

Terms were therefore agreed with all four groups of investors by late April and the total cost of the share acquisition was £36,270. Of this amount a maximum of £27,500 could be obtained from the bank and the remainder, apart from any internal resources of the company which could be used, had to come from the new investor. The new investor was Myles Cooper, director of Stavert Zigomala, a leading Manchester company of textile merchants. His investment in Dorman and Smith Limited served two purposes. Firstly, he was seeking a suitable employment for his son, Bruce Cooper who, it was hoped, after suitable training and experience, would assume a senior management position in the company. In fact he was to be trained in the drawing office and later to become a director of the company. Secondly, it was a financial investment for which he wished to have sufficient safeguards to protect his investment and dividend payments. He had no wish to participate actively in the running of the company. On this basis Myles Cooper agreed to purchase 7,000 cumulative preference shares for £5,000 in the new company. Of the 14,000 ordinary shares to be issued, one was to be held by Haworth and the remainder by Thomas Atherton, the chairman and managing director of the new company.

On 5 May 1937 the purchase of all outstanding shares was completed and on 8 May Dorman and Smith Limited went into voluntary liquidation. A new company, Dorman and Smith (1937) Limited was brought into being on 1 May and on 10 May acquired the assets of the former company. The inclusion of '1937' was required for the purpose of legal distinction from the former company and this was dropped shortly afterwards.

Thus, in the space of just four months, the company of Dorman and Smith Limited had been reconstructed and a new majority shareholder, company chairman and managing director established on the basis of an initial expenditure of one hundred pounds. But for this a price had to be paid. Twenty-seven and a half thousand pounds were owed to the bank and had to be repaid at a considerable rate, whilst the company now possessed a new preference stockholder holding significant powers to protect his interests. The future prospects of the company now rested in different hands and it is fitting that this chapter in its history should conclude with an appreciation of its new chairman from the man who had controlled its fortunes during most of this difficult period. In a farewell letter to Thomas Atherton at the end of August 1937, H. G. Baggs records 'my grateful appreciation of your loyal help during our period of difficulty: and my high appreciation of your own good judgment and great courage and enterprise in business'. The exercise of those qualities in the renewed expansion of the company is the subject of the next chapter.

TABLE 9 FINANCIAL RESULTS OF DORMAN AND SMITH LIMITED AND GENERAL ECONOMIC INDICATORS, 1914–37

Year ending 31 March	Dorman and Smith Limited: Total net assets £ *thousands*	Total sales £ *thousands*	Net profit (after management payments) £	Average rate of dividend interest paid by electrical manufacturing companies (per cent)	Electricity sales (*millions BTUs*)
1915	40	45	2,861	6.1	1,635
1916	44	54	4,182	6.7	1,694
1917	46	62	4,097	6.2	2,001
1918	41	66	3,275	6.6	2,367
1919	48	78	4,788	7.8	2,716
1920	52	85	4,887	8.1	3,079
1921	63	111	13,907	10.0	3,512
1922	61	72	5,359	7.8	3,121
1923	49	55	1,708	7.4	3,762
1924	47	60	1,864	7.1	4,468
1925	48	76	2,905	7.4	5,097
1926	54	85	8,689	7.8	5,606
1927	54	79	7,048	7.7	5,868
1928	54	70	4,567	7.7	7,003
1929	53	70	3,731	8.4	7,800
1930	50	67	3,038	9.0	8,666
1931	41	48	– 2,874	8.6	9,073
1932	37	38	– 4,079	7.3	9,501
1933	36	30	– 5,383	6.6	10,210
1934	37	35	– 172	6.7	11,467
1935	36	37	– 365	7.3	13,030
1936	37	42	303	8.5	15,049
1937	38	52	4,797	9.9	not available

Sources: Dorman and Smith Limited accounts, and *Garcké's Manuals*.

PLATE 15. Switchboard supplied to John Brown and Company for installation in SS *Arnhem* about 1948, typical of Dorman and Smith marine switchgear

PLATE 16. Dorman Smith switchgear was installed throughout the Houses of Parliament 1947–9. This dc switchboard operates lift services

CHAPTER FOUR

Thomas Atherton and the Revival of the Company 1937–48

The new company, Dorman and Smith (1937) Limited, was registered as a private limited company on 8 May 1937 with its office at the Ordsal Works in Salford. The directors were Thomas Atherton, the architect and driving force of the new company; Major Richard Amberton, the London agent of the old company who joined the new board at Atherton's request; John Noel Haworth, an accountant appointed to represent outside interests, who was also auditor to the company, a joint responsibility which was legal then, but is no longer so; and H. G. Baggs. The full board met for the first time on 14 May and appointed Thomas Atherton as chairman and managing director. He was unquestionably in command, for Baggs's appointment as director and technical adviser was for a specified period of only five months – virtually a courtesy appointment but a very generously paid one since he was to receive £650 for occupying this position from May to September inclusive. This represented an annual rate appreciably higher than that of the new chairman, whose salary was set at £1,000 for the first two years, and £1,500 a year thereafter. Haworth's fee was set at sixty guineas a year, and Amberton retained the London agency and was voted a fee of three guineas for each directors' meeting and shareholders' meeting he attended. Myles Cooper, though holding the entire preference capital, was purely a sleeping partner as he wished, without a seat on the board.

For the first few days, Mr John Henry Bull served as company secretary, but Atherton had already approached Mr Robert Dale in March, told him of the negotiations that were in progress and asked whether he would accept the post of secretary to the new company. Dale agreed and, on 14 May at the age of thirty, became its secretary. He had joined Dorman and Smith as general office manager in 1934 and thus knew Thomas Atherton well. He remained secretary until his retirement in 1967 and he was also appointed a director of the subsidiary company, Switchgear Units, on its formation in 1940.

It is clear from the previous chapter that Thomas Atherton was risking everything on his judgement that the company could do well under his leadership, and emerge from the doldrums into which it had drifted for the past decade. The memorandum of association, which declares the areas of interest of a company and is usually framed very broadly so that companies can expand without administrative difficulty, reflected both the ambition of the new chairman and his view that the rearmament programme would expand the scope of the company's operations; for among the possible manufactures listed in the memorandum were guns, projectiles, aeroplanes and seaplanes.

In the articles of association, however, there were some constraints upon the chairman. The consent of the holders of three-quarters of the preference capital was required for any increase in the share capital, or for any increase in any director's remuneration above £1,000 during the first two years of the company's life, or above £1,500. In essence, this gave Myles Cooper an upper limit on the salary and commission of the chairman. However, any new shares that might be issued in the future had to be offered to members in proportion to the existing distribution of shares. Thus, Thomas Atherton was securely in control, but Myles Cooper had some safeguards for his own sizeable investment.

Those who worked with Atherton at this time testify to his enormous energy and determination. Robert Dale, then secretary, recalled in 1970:

> I won't pretend that he didn't have his periods of worry, but he was a man of terrific energy – in fact, there were times when I begged him to slow up, because I wanted to slow up.

The news of his accession to the chairmanship, and the attendant financial burden that he bore came as something of a surprise to the older members of his family: here was young Thomas Atherton, not yet forty, with a wife and two young sons, staking everything on his ability to make Dorman and Smith Limited into a much more profitable company. A letter from an uncle in May 1937 shows the surprise he caused in the family:

> Dear Tom,
>
> Your Mother has told me that you have taken over Messrs Dorman and Smith's, and to say that the news has somewhat staggered me is only the truth. It would appear to a simple chap like myself that this is a very big thing, but I do very sincerely wish you every possible success.
>
> I should fancy the re-armament programme will help you very considerably as such a colossal programme is bound to be reflected in a business such as yours.
>
> When one comes to think back 30 years and remember you as you were then, your progress has been wonderful. I hope this task will prove to be well within your power. Good luck to you, Tom, and may success be yours.

However, while the family marvelled at his pluck, Atherton was already considering his future strategy. We gain an insight into this from a letter to his bankers, actually written during the negotiations of April 1937. Most of the letter was taken up with an explanation of the new company's prospects, but at the end there was a revealing paragraph:

> There is one other point. If Mr Cooper takes up participating preference shares and I enter into the proposed agreement with him, I shall be in his hands with regard to the running of the company. As you know, it is at the back of my mind to sell the Company to the public in a few years if things go as I expect them to do. Capital appreciation probably

> is not as important to Mr Cooper as it is to me; but I take it that he would not be unreasonable if the opportunity occurred to float the Company at considerable profit to both of us.

Although the war intervened to postpone these plans, ultimately the intention was realised when preference shares in Dorman and Smith were offered to the public in 1948, and ordinary shares in 1962.

Business conditions on Atherton's succession

Clearly, Atherton was confident that he could succeed, as was mentioned in the preceding chapter. Even during the transitional period while the new company was being formed, there were immediate grounds for optimism. For example, orders received in the first seventeen days of April 1937 were as much as those received during the whole of April 1936. Sales for the full month of April 1937 were over £5,000 which was about sixteen per cent higher than the monthly rate for the financial year 1936–7 just ended, and forty-five per cent higher than those for 1935–6.

On the face of it, therefore, it would seem that the takeover was arranged at a very opportune time; but in fact 1937 was, in some senses, the crest of a wave and general economic conditions became slightly less favourable in 1938, though still very much better than during the worst years of the depression.[1] To see how economic conditions impinge upon the company, one must look at the pattern of investment expenditure at the national level. The products of the company were essentially related to investment expenditure by the ultimate customers, such as builders installing electrical equipment or factories extending or modernising their facilities, for although they included a number of simple domestic lampholders that the public might buy, much the greater part of the company's catalogue and sales in 1937 comprised more specialised items, such as watertight or acid proof lightfittings, bulkhead lightfittings, switchgear, circuit

[1] D. H. Aldcroft puts the turning point at September 1937: see his *The Inter-War Economy: Britain, 1919–1939*, Batsford, London, 1970, p. 28.

breakers and distribution boards. Hence, while electricity consumption was rising steadily this did not mean that Dorman and Smith's sales followed suit; they were likely to be much more erratic and follow the investment cycle of the economy. In fact, in 1938 electrical manufacturing suffered a minor setback, and unemployment among workers in the sector manufacturing electric cable, apparatus, lamps, etc rose from five per cent in 1937 to 8.1 per cent in 1938.[2] Wage rates, however, rose during the same period by nearly four per cent in the engineering industries,[3] increasing the pressure on employers while demand flagged. New investment in dwellings fell between 1937 and 1938 by just over one per cent in real terms, but investment in non-residential buildings fell by about 5.2 per cent.[4]

Some of these problems were reflected in the company's experience. Sales for the year ending March 1938 were fourteen per cent higher than those for the previous year, though as we saw earlier they had been running at sixteen per cent above the old level. However, by the standards of the industry it was a very creditable performance for the new management, and prompts one to ask what policies Thomas Atherton introduced in order to achieve it.

Changes in policy

The new chairman and managing director made a number of changes. In isolation they were not startling, but they must be judged against the background of the previous regime, and in the light of the company's fortunes since the change of management.

One of the weaknesses of Baggs's management had been his impatience with subordinates. His real passion had been for good engineering, rather at the expense of human relationships within the works: as well as the changes in senior management mentioned in the previous chapter, there were even more frequent changes

[2] B. R. Mitchell and P. Deane, *Abstract of British Historical Statistics*, Cambridge University Press, 1962, p. 67.

[3] *Ibid.*, p. 351.

[4] C. H. Feinstein, *Domestic Capital Formation in the United Kingdom 1920–38*, Cambridge University Press, 1965, p. 38.

lower down the hierarchy. Robert Dale, who served as general office manager under Baggs, recalled:

> During the years 1934 to 1937, particularly in the drawing office, there were changes almost every week. We had six or seven senior draughtsmen. Also the buying department was in-out, in-out – some of the people were completely nervous wrecks.

Baggs was little more diplomatic in his treatment of some of the firm's customers. Thomas Atherton recalled in 1970:

> Baggs was a very good engineer, but had no feeling for customers. He didn't produce a design, he went for one-offs, and what he said went. I remember a customer who said he wanted a common-or-garden switch, and Baggs replied that he didn't make switches for the music-hall stage . . . I saw people going by (my office), Baggs appointed them, and Baggs decided that they ought to go. I didn't know whether I was next, as I had succeeded Wilkins, whom Baggs dismissed even though he had been very friendly with him in his youth.

Atherton quickly re-established two vital principles: he determined to apply commercial criteria in running the company, and he quickly built up good relations with his subordinates, establishing a reputation for toughness and determination, tempered by fairness in all his dealings. Dale recalled that the climate changed overnight, and the Dorman and Smith company became a business with a leadership by someone who knew what he wanted to do. Atherton's accession was felt throughout the company, as two of his changes show. The first, a minor but amusing alteration, concerns the bell at the 'Inquiry' window in the foyer of the Ordsal Works. For years there had been a spring-loaded bell operated by a lavatory chain. On the Monday of the first week of Atherton's management, he turned to Dale and said, 'I'm having that damned thing taken out.' It was duly replaced by a much more appropriate electric push bell.

The second reform was of more direct concern to the workers at Ordsall. Working hours on Saturdays were from 9 am until 12.45 pm. But little was done in the last three-quarters of an hour, and Atherton had always found it inconvenient as it made him late for lunch at his home in Southport; so to everybody's satisfaction the closing time on Saturdays was brought forward to mid-day.

The rapid turnover of staff ceased under the new management, so that employees felt much more secure and confident. New staff were engaged in various capacities. On 1 June 1937 the company signed a service agreement with Bruce Lusk Cooper, the son of Myles Cooper. Myles as a major shareholder was anxious that his son should enter the business, though Atherton was rather perturbed about the proposed arrangement at first. However, it was not a case of unjustifiable nepotism, for the younger Cooper learned the business well and later joined the board and became managing director of DS Plugs. Eventually he was to become vice-chairman of Dorman Smith Holdings Limited, a post he still occupies. At a board meeting in July 1937, Atherton showed his appreciation of the need to improve the marketing facilities of the company by proposing the appointment of four sales representatives. His co-directors agreed that two should be appointed as soon as possible and that two others would be considered at the end of the financial year in March 1938. In August 1937, J. E. Adey became northern sales representative, and several applications came in for the post of Scottish representative, which was eventually filled on 11 October. Baggs formally retired and was presented with an engraved silver plate in recognition of his forty-five years spent with the company, more than half of them as a director.

Baggs's retirement emphasised the need for a designer, for Atherton had no background on the technical side (though he soon showed himself to be receptive in this area). During the summer of 1937, the London agent, Major Richard Amberton, interviewed a number of applicants for the post of designer. Amberton was an extremely important connection of the company, as their correspondence shows. Fifteen years older than Atherton, he had been concerned in electrical engineering since he left school, and was a member of the Institution of Electrical Engineers. He widened his experience by working in the United States, and after being invalided out of the Royal Engineers in the first world war he directed various sections of the Department of Engineering at the Ministry of Munitions for two years. Later he established the consulting firm of Amberton and Partners, which became London agents of Dorman and Smith at the end of 1934. Physically he was a most striking figure – unusually tall, monocled and extremely well dressed – and he was a keen golfer and bridge player as well as a very prominent Mason.[5] On 4 October 1937 Thomas Smalley began work as the company's designer, joining them from Salford Electrical Instruments Company.

Another member of the company who advanced after the changes in 1937 was an accountant, Clarence Bickerton. Both he and Smalley prospered in the new company and together with the secretary, Dale, were appointed to the board of the subsidiary company, Switchgear Units, when it was formed in 1940. The atmosphere of amity was in sharp contrast with the strained relations that existed before 1937, and the credit for it undoubtedly belongs to Thomas Atherton, with his ability to select the right men, and work with them.

Smalley proved a capable designer. He was often called into board meetings to report on the progress of new designs, especially the Lancastrian range of switches, and his salary was comparatively high. For example, new three-year service agreements were signed in January 1938 between the company and Smalley, Dale and Bickerton, each one specifying a basic salary plus an element of commission on sales. Dale and Bickerton were each paid £312 a year plus one-sixteenth per cent on net monthly sales and one-quarter per cent on any excess of annual sales over £65,000; Smalley received £450, plus one-quarter per cent on sales in excess of £65,000 – an arrangement which almost guaranteed him £100 a year more than his two colleagues and insulated him more from the risk of a decline in turnover. Early in 1938

[5] Most of this information is contained in an obituary published in the *Journal of the Institution of Electrical Engineers*, October 1958.

Smalley and the company applied jointly for two patents on improvements to electric switches and circuit breakers. By mid-1938 Smalley succeeded Mr G. H. Platt as works manager, and was awarded an extra one-eighth per cent on net monthly sales. He also applied, jointly with Thomas Atherton, for a patent on improvements to machine joint boxes for electric switches and circuit breakers.

There are two significant implications that follow from this account. One was that the company was again striving for a technical edge in the market, whereas Baggs in his latter days had tended to scorn patenting; the other was that Atherton, by putting his senior executives on service agreements that directly linked their pay to the company's sales, and by guaranteeing them against capricious dismissal, ensured a community of interest in the success of Dorman and Smith.[6] This was also true at the shop floor level, for in October 1937 girl machinists and boys operating capstan lathes were put on piece rates to speed up production. In the first month, output went up by thirty-five per cent, and wages paid by twelve per cent. Later in 1938 piecework was extended further in the factory. Another improvement for the employees was the establishment of a holiday fund in early 1938, following the agreement between the Engineering and Allied Employers' National Federation and the Engineering Joint Trades Movement.

The outcome of all these changes was a clear boost to both morale and efficiency. But Atherton realised that it was equally important to cultivate successful relations with customers, and in March 1938, the Admiralty Inspectorate and the Air Ministry Inspectorate visited the Ordsal Works. They both expressed their entire satisfaction with it, and put the company on their approved list of suppliers. Many of the products of the company were directly relevant to military requirements such as ships' lighting fittings, and the switchgear was, of course, vital in the construction of munitions factories; however, the visit had a deeper

[6] Another employee who gained a direct stake in the company was Bruce Cooper, to whom Myles Cooper transferred 1,000 of his 7,000 preference shares in April 1938.

significance in that it effectively made the company eligible for the specialised contract work that the war effort necessitated. Orders were soon received and in March 1939 Thomas Atherton gained assurances from the District Bank that they would advance up to £3,000 to help finance a contract for the supply of switchgear to the Royal Ordnance Factory at Glascoed, Monmouthshire; this precaution was necessary because civilian business was also quite buoyant so that the company had to bear the cost of a significant increase in stocks and work in progress. Among the important civilian contracts secured at this time was one for the supply of substation switchgear to Salford Corporation.

Financial progress

It is not easy to assess the financial performance of the reconstituted firm against its results under the previous management; while sales can be readily compared, the rate of profit cannot because at the formation of the new company in 1937 the assets of the company were revalued, as the figures in earlier balance sheets had been extremely conservative and the bank had to be reassured that it was financing something more substantial than the purchase of goodwill. Thus, while there were no additions to land, buildings, plant, fixtures or fittings during the first year of the new company, the amount attributable to these items in the balance sheet on 31 March 1938 was £25,349 compared with only £11,778 in the old balance sheet.

Profits in 1937–8 showed a continuing improvement, though it was not all accruing to the shareholders, as it would have done before the change of management. There was now £980 to pay to the bank on the mortgage debenture. There had also been preliminary expenses connected with the establishment of the new company, a payment of £650 to H. G. Baggs, and a depreciation provision in 1937–8 of £1,451, whereas in 1936–7 there had been no comparable charge, presumably because the assets had been so undervalued.

The figures in Table 10 show the rapid progress in the sales and profits of the new company.

TABLE 10

GROWTH OF DORMAN AND SMITH SALES AND PROFITS, 1936–7 TO 1940–41

Year	Sales £	Profit £
1936–7 (old company)	51,918	4,797
1937–8	59,145	5,691
1938–9	66,383	4,576
1939–40	107,477	10,103
1940–41	166,388	10,370

Thomas Atherton had two clear aims during this period. The company had been financed largely by the District Bank, which held the mortgage debenture, and the Cooper family, which held the preference capital. Practically all the ordinary capital was held by Atherton, and though the revaluation of assets gave his equity an imputed value of some £18,000, his prosperity rested upon the realisation of his two aims. One was to repay the mortgage debenture as quickly as possible, for every pound repaid effectively increased his proportionate share of the business. The other was to build up the profits of the company, because as the holder of all the ordinary shares, once the preference holders were paid, two-thirds of the profit were his. In financial parlance, Atherton was running a very highly geared company; the distinguishing features of such companies are that they reward those who run them extremely well, so long as the business prospers – but if business worsens and profits contract, the interest payments can impoverish the ordinary shareholders.

However, the company did prosper and earn profit, and regular repayments were made to the bank. In the first financial year, the debenture was reduced from £27,500 to £25,000 and in the second year to £22,000. However, the pressing needs of war time expansion prompted the company to open other mortgage debentures, while repaying the original. The preference dividend due to Myles Cooper was not paid in the first two years, which naturally caused him some concern. In fact Atherton had wanted to pay the preference dividend but the bank, which had prior claim on the debenture holder, had asked that the preference dividend be withheld for the first two years of the new company's life. The preference shares were cumulative, and any arrears in the dividend payments had to be made good in the future.[7] Atherton wrote to reassure Myles Cooper in July 1938 that although he had been obliged to pass Cooper's dividend, the company was in good health:

> We have repaid £5,000 debentures to the bank. This time last year we had a balance of orders in hand amounting to £8,250. Today the balance is £13,700. This notwithstanding that our competitors have experienced a recession in the trade.

In October 1939, Major Amberton wrote from London to Thomas Atherton urging that he should get the agreement of the bank to allow him to pay off the arrears of preference dividend. Amberton emphasised that the sum involved was small; that the confidence of the preference shareholders was affected; that customers might lose confidence in the company if they discovered that its preference dividends were unpaid; and that Atherton's own wishes and peace of mind would be realised, and he would have greater freedom of policy if there were no arrears. In the event, the arrears were paid in 1940 and 1941 to the relief of both Atherton and Cooper.

Rearmament and the second world war

We have already mentioned that rearmament brought important government business to the company. The business was not

[7] Item 7(5) of the articles of association safeguarded the preference shareholders by limiting the salary of any director to £1,000 as long as their dividends were in arrears.

fortuitous, for Atherton had taken pains to establish good contacts with the armed services, and was entertaining some admirals to lunch on the day war broke out.

Although after the Munich Agreement of 29 September 1938, the general public in Britain believed war had been averted – and Dorman and Smith shelved their plans for an air raid shelter at the works – the government began to expand the rearmament programme.

A historian of the period noted:[8]

> In 1936 and 1937, the schemes of expansion proposed by the Services, were considerably whittled down to save money. Only in the course of 1938 did the government come round to the view that other aspects of defence must outweigh considerations of financial economy . . . After the Munich agreement, the government at last decided that Britain should prepare an army for continental service and expand munitions production to meet its needs, despite the heavy costs involved.

The military contracts became increasingly important to Dorman and Smith, as they did to other firms at the time.

Indeed demand became very strong, as the following letter from Atherton to Amberton in September 1938 shows; it also shows the pressure upon the business in seeking to meet the demand:

> We have at the present time two unoccupied rooms in the works of a total space of 10,500 square feet, of which you are aware. In addition we could provide accommodation in existing workshops for at least 100 more workers. This is an absolute *minimum*.

[8] W. Ashworth, *Contracts and Finance* (History of the Second World War), H.M.S.O., London, 1953, p. 10.

You will see therefore that we can increase our output very considerably simply by filling up space without putting down any new buildings. In addition we are prepared to put on a night shift in certain departments if we can get labour and I think that just at the moment we can get it, although if we are to wait a month or two for a decision the labour market will not be so easy.

We should have to consider very seriously indeed before we put down any expensive extensions *and we should require some sort of financial guarantee from the Office of Works*, on the lines of the Aircraft Factories.

Before we can go much further into the matter we should have to have more definite proposals . . . meantime I should like you to assure Mr Chappell (*ie* the Office of Works representative) that we are willing and anxious to do everything we possibly can to help them in this emergency.

As mentioned to you over the telephone, we are prepared to take the Bishopton job at cost plus an agreed percentage. We are agreeable to guarantee deliveries – if necessary other work would have to go by the board, but I do not think this would be necessary.

Mr Chappell and his assistants are well aware of the quality of our products and, if they will permit us to go ahead, we shall maintain that quality and give them a really first-class job. We have probably more experience in this type of work than any other company in the country.

P.S. I have just run through this Bishopton job again. If they would place it with us on our standard NC or, preferably, Lancastrian gear we would guarantee to deliver every switchboard to them before the end of December. If Mr Chappell does not move quickly he will probably find us booked by one of the other departments by next week.

War time problems

Atherton's earlier judgement that rearmament would create a lot of work for the company was thus amply borne out. The minutes of the directors' meeting in November 1939 record 'greatly increased orders' and mention two proposals to enlarge the Hampson Street Works. However, the actual outbreak of war raised some major problems. So long as the government demand was for switchgear and associated electrical equipment, Dorman and Smith could invest in any extra facilities required without any qualms about their ultimate usefulness. But proposals that they manufacture actual munitions placed them in a dilemma because much of the capital equipment that they would need to buy in could be useless when the war ended. Amberton sent a memorandum to Atherton at the end of October 1939 concerning a proposal that the company should manufacture artillery shells. He emphasised that most of the plant, space and skill devoted to shells would be useless when the war ended, whereas switchgear plant, or plant for other mechanisms outside the company's traditional field would be a good basis for peace time business. Ideally he hoped that war time activities might help to contribute to building over the spare ground at the works, modernising and extending the plant, and acquiring a plating and polishing plant which would help to improve the finish of the standard Dorman and Smith products.

This probably seemed a much more pressing issue than it does in retrospect, for it was the time of the *Sitzkrieg* or 'phoney war', before the German invasions of Scandinavia or France. Amberton's optimistic postscript at the end of the memorandum – 'I am beginning to doubt if the war will last as long as two years' – was no more than a reflection of the public mood that prevailed.[9]

In fact the company succeeded in avoiding too heavy a commitment to narrowly specialist military production, and their skills were valuably employed in making components for ASDIC, the submarine detection apparatus, and for the RAF, which

[9] For an account of conditions and attitudes at the time, see A. J. P. Taylor, *English History 1914–1945*, Oxford University Press, 1965, pp. 453–466.

specified tolerances as close as one ten-thousandth of an inch. The knowledge of the company's designer, Smalley, proved useful in the ASDIC work, and the company also made control boards for radar sets. The senior executives were extremely busy, and though nominal hours of work were from 9 am to 5.30 pm, Robert Dale, the company secretary, usually counted himself lucky to be at his home in Gatley by 9 pm.

If Dorman and Smith had problems in financing their war effort, they pale into insignificance beside those of the government. In their relations with companies, the government had to give them sufficient inducement or assistance to persuade them into the war effort, but not be so lax in its policies that employers could make enormous profits from war contracts – for the memory of profiteering in the 1914–18 war was still vivid. Indeed, Mr Chamberlain pledged the government to 'taking the profit out of war'.[10] An armaments profit duty was imposed in the summer of 1939 to prevent armament firms earning unreasonable profits. However, many firms outside the armaments industry, such as Dorman and Smith, were receiving orders directly connected with the war effort, so in the September budget an excess profits tax (EPT) was imposed, back dated to 1 April 1939. A tax of sixty per cent was levied on any profit over and above a specified pre-war standard.[11] Firms, like Dorman and Smith, which had come into existence since 1936, were simply allowed eight per cent return on their capital employed; fortunately for the company this was defined to include borrowed capital. Thus three-fifths of all profits above the eight per cent level went to the new tax, and in May 1940 the rate was raised from sixty to one hundred per cent, thus confiscating *all* profits above eight per cent. This swingeing rate was a political move, and prompted much argument that it destroyed enterprise and pressure for efficiency. Here the definitive history includes a general statement which is particularly relevant to Dorman and Smith:[12]

[10] House of Commons Debates, Vol. 346, Col. 1350.

[11] Fuller details are given by R. S. Sayers, *Financial Policy 1939–1945* (History of the Second World War), H.M.S.O., London, 1956, p. 29.

[12] *Ibid.*, p. 86. The quotation included is from a memorandum by Lord Stamp.

PLATE 17. The fused plug developed by Dorman and Smith, using the novel *Alorite* fused pin, introduced in 1945

PLATE 18. TS switch, designed by Thomas Smalley and introduced in 1951

PLATE 19. The Dorman System of modular switchboard construction, introduced in 1954

PLATE 20. The Cubicon system of switchboard construction, introduced in 1955

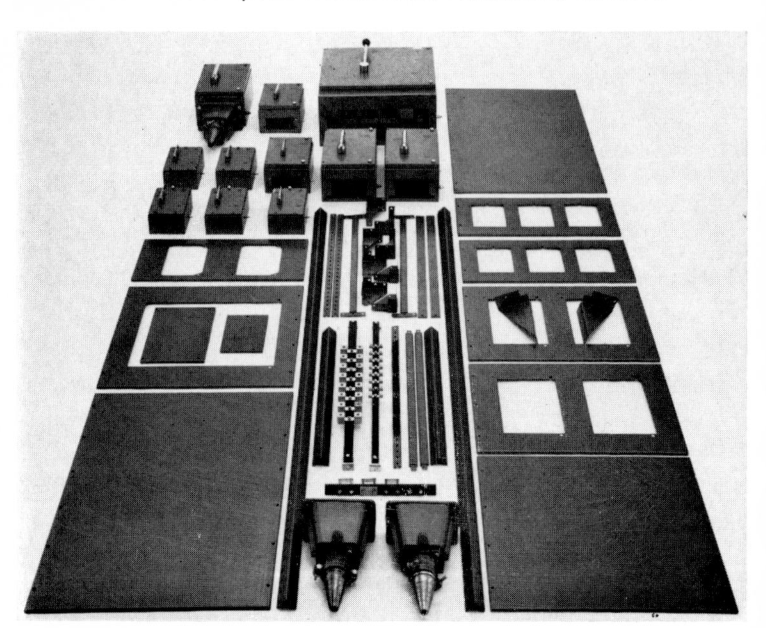

PLATE 21. Components of Cubicon section of switchboard supplied to customers for assembly on site

> These arguments were of particular force in new and expanding businesses, where also the availability of profits as a source of capital expansion was of most importance. The danger was, in fact, not so much that of glaring cases of a refusal to undertake work of national importance that is clearly within a manufacturer's capacity. The danger is more subtle; it is likely to take such forms as a gradual decline of zeal, and energy and enterprise...

The excess profits tax was a severe blow to Thomas Atherton, as his strategy for building up the business and his stake in it turned upon earning high profits to repay the mortgage debenture. These profits were now pre-empted by the new tax; even when the government yielded to public pressure against the tax and announced in 1941 that twenty per cent of excess profits tax paid would be refunded after the war, it did not help Atherton, because it still tied up the profits. With EPT at one hundred per cent Dorman and Smith could not hope to meet interest and debenture repayments. On the basis of the 1939 accounts for example, the maximum amount of profit that would be left over for annual dividends on the ordinary shares and repayment of the debenture would be about £2,650. Atherton conferred with Haworth, the accountant, who was a co-director of Dorman and Smith, and C. J. Hall, his solicitor, and sorted out a scheme which met with the approval of the bank.

The formation of Switchgear Units Limited

Switchgear Units Limited was incorporated on 26 August 1940. The rationale of the company was to enable Thomas Atherton and Bruce Cooper to increase the income that excess profits tax had virtually frozen. Dorman and Smith Limited were to make loans to Atherton and Cooper, who would then give cheques to nominee shareholders, enabling them to buy shares. The share distribution was as follows:

Robert Dale, chairman	1
Clarence Bickerton, director	1

Thomas Smalley, director	1
Myles Lusk Cooper, director	251
Bickerton, Smalley and Dale	246 jointly
Total	500

Thus neither Atherton nor Bruce Cooper had any legal shareholding in the new company, but there was an unwritten understanding that the shares were not the property of the declared holders, but really belonged to Atherton and Bruce Cooper, who signed service agreements with Switchgear Units giving them each £250 a year salary, plus expenses and five per cent commission on turnover. Atherton was sales manager and Cooper works manager. All their income received under the service agreements was to be paid into a joint account and, after tax, devoted to the purchase of shares in Dorman and Smith in the ratio two-thirds to Atherton and one-third to Cooper. The arrangement made it possible for Atherton and Cooper to earn fairly large sums of money without incurring excess profits tax; since neither was a shareholder, it could not be deemed a subsidiary of Dorman and Smith. Instead of all growth occurring in Dorman and Smith, with no prospect of profits growing in step with turnover, business could be expanded in Switchgear Units, where the income of Atherton and Cooper was not bottled up by excess profits tax but was related to the amount of business they did. It was a war time expedient to get round a problem which Atherton could not have foreseen when he took over Dorman and Smith in 1937. It was not necessarily expected to be permanent, as it was recorded in a personal memorandum by Atherton that 'when the company is eventually wound up, the assets (after repayment of loan) to be applied to the purchase of (Dorman and Smith) shares 2/3rds to TA, 1/3rd to BLC.'

However, Switchgear Units was not a facade. It began production on 28 October 1940, in rented premises at Lizzie Street, in Pendleton, which was a little over a mile from the Ordsal Works, and in March 1941 the Breightmet Works at Bury Road, Bolton, was rented to cope with the growth of Admiralty business. The growth of Switchgear Units' activity is shown in the following table, which illustrates its importance relative to that of Dorman and Smith Limited, and the degree to which it was geared to the war effort, for Switchgear Units showed a decline as the war ended, whereas Dorman and Smith did not because it was better adapted to peace time demands.

TABLE 11

SALES OF DORMAN AND SMITH, AND SWITCHGEAR UNITS, 1938–9 TO 1945–6

Year	Dorman and Smith Limited	Switchgear Units Limited
1938–9	66,383	—
1939–40	107,477	—
1940–1	166,388	24,728*
1941–2	165,812	93,823
1942–3	188,999	106,397
1943–4	222,270	140,968
1944–5	244,556	128,629
1945–6	245,644	56,261

* Seven months' operations only.

The success of Switchgear Units was rather more than Atherton expected, and he and Cooper reduced their commission from five to three per cent; but its growth proceeded in an unforeseen way. Its original intention – to help offset the excess profits tax which bore more heavily on Dorman and Smith than most companies – was partly frustrated. The Inland Revenue, which ruled in 1943 that the company was separate from Dorman and Smith for tax purposes, nonetheless effectively prevented it from making any significant profit[13] or paying B. L. Cooper more than £250 under the service agreement, and disallowed Thomas Atherton anything. An appeal brought only limited concessions.

[13] For example, in 1943–4 profits after tax were £573, or 0.4 per cent on turnover.

At the end of the war, the company was not wound up, and Dale, Bickerton, Cooper and Smalley retained their shares, though relinquishing their directorships. Instead, the plug and socket section of Dorman and Smith was transferred to Switchgear Units, and large sums of capital were invested by the Atherton and Cooper families, and small quantities of shares were given to Richard Amberton, Dale, Smalley, Bickerton, Jack Lund and Charles Nesbitt – the two last named later to become chief designer and sales manager respectively. The name of Switchgear Units was changed to DS Plugs Limited, which was incorporated as a public company in 1946 although its shares were not quoted on the Stock Exchange. What had begun as a temporary war time enterprise became the cornerstone of a very significant peace time business, to which we shall return presently.

Problems of war time finance

The unusually heavy taxation of war time made it difficult to earn enough profits to buy the machinery necessary to expand production. This problem is mentioned by Thomas Atherton in a letter dated 25 January 1940 to A. E. H. Pew, the principal technical officer of HMS *Osprey*, a shore establishment at Portland Naval Dockyard. At the time, the company held contracts to supply switchgear, plugs and sockets worth £24,300 to the Admiralty, but taxes would reduce the profit on these contracts to £600. The letter notes:

> The margin of £600 which we can expect to receive out of these orders is not adequate to purchase the amount of plant which is necessary, especially when it is realised that second-hand machine tools have more than doubled in price since September, and that new machine tools are unpurchasable. The contractor is also faced with the difficulty that at the end of this present war when the Ministry of Supply throw their surplus tools on to the market, the machines which he buys now will not be worth more than 25% of their cost, and

> most of them will be useless to him because Government orders will fall to their peacetime levels.

Atherton went on to say that he had discussed the problem with the Admiralty electrical engineer in Manchester, who had advised him to consult Pew rather than the Admiralty at Bath. He requested that the Admiralty should supply the necessary machines and reclaim them at the end of the war.

In fact, such arrangements were common practice. The Admiralty had used capital assistance schemes, as they were called, since 1937, and they were much the commonest form of assistance or incentive;[14] they accounted for eighty-one per cent of Admiralty contributions to fixed capital during the war. Under these provisions, the Admiralty provided a variety of machine tools for both Dorman and Smith and Switchgear Units, though it should be added that the rate of profit allowed on the use of these machines was, rightly, kept low. The Admiralty also paid for an air raid shelter and wooden canteen at the Breightmet Works of Switchgear Units. At the end of the war, the Admiralty sold a number of machines to the company and these formed a useful basis for peace time production.

Operations in war time

The war predictably drained the company of much of its manpower, at the very time when it needed more people. Many women were taken on until the whole of the office staff were female. The directors and senior staff members whose age or reserved occupational status excluded them from military service volunteered for ambulance or air raid precautions work. Others were on active service: Myles L. Cooper, a director of Switchgear Units, served four and a half years in the merchant navy and was torpedoed in *Laplace*, a cargo ship of the Lamport and Holt Line. He spent eight days in a lifeboat off the Cape of Good Hope. Stanley Royds, who began as a cost clerk under Mr Bickerton and has since risen to become contract sales manager, was called up into the RAF and completed several tours of duty with

[4] For further details of these see Ashworth, *op. cit.*, pp. 204–214.

Bomber Command. Clearly a full list is beyond the scope of this work.

Those remaining worked long hours. In 1942, the Ministry of Labour issued an instruction that the standard working week should be forty-six hours, but usually it was appreciably longer than this.

Fortunately the company was not very seriously affected by enemy bombing and no lives were lost. An air raid on the night of 22 December 1940 caused some damage at the Ordsal Works and the office staff and some production workers were moved temporarily to the Pendleton Works. The following night another bomb fell near the Ordsal Works but failed to detonate; however, the works had to be closed by order of the civil authorities. Production ceased, and the works was not reopened until 3 January 1941; the stoppage was estimated to have cost the company £4,026. There was more damage to the Ordsal Works and its air raid shelter on the night of 1 June 1941 but no significant interruption of production.

One of the lasting innovations made by the company during the second world war was its pension scheme, which was first mooted at the end of 1943 and introduced in 1944. It was open to men on the staff between the ages of twenty-one and fifty-five, and women between twenty-one and fifty, and paid on retirement thirty per cent of the salary received ten years before retirement. At that time inflation was much less severe than it is in the nineteen seventies.

A more temporary innovation was the manufacture of machine tools, for at one period supplies were so scarce and the resultant bottlenecks to production so severe that Dorman and Smith designed and manufactured a small number of their own lathes.

Policy for peace time: development of the fused plug

Throughout the war, both the companies remained virtually flat out on production, but still found time to produce a number of innovations. In August 1941, for example, three applications for patents were submitted covering improvements to fuses, a floating contact fuse and an insulated distribution box. In 1942 some applications were also made for patents in Australia, New Zealand and South Africa. Thomas Smalley showed the board an improved design of contact breaker at the end of 1942, and early in 1943 it was demonstrated in London, Liverpool and Glasgow. In March 1943 an application was made to patent improvements to electrical connector plugs, and in the following month came an application of great significance to Dorman and Smith, for 'improvements in and relating to high-rupturing-capacity cartridge fuses'. The significance of these lay in the development of the fused plug, which was probably the most original and exciting of the company's new products for three or four decades. Perhaps the most unusual aspect of it is that the plug was conceived not by somebody with formal technical qualifications, but by the chairman, Thomas Atherton. The inspiration for the plug came when he found the maid at his house in Southport trying to join two lengths of cable together because a fuse had blown.

> I thought this was all wrong, so I designed a plug where everything was in the eye of the beholder; you shouldn't have to take a plug to pieces to get at the fuse. I was very technical at that time. I put the first experiments in hand with regard to the cartridge fuse; I'd got to do because I hadn't got much money – I couldn't afford to take extra designers on.

The domestic ring circuit method of wiring, commonly known as the 'ring main' today, had been advocated by a committee appointed in June 1942 by the Institution of Electrical Engineers. Its findings were published in 1944 as Appendix 4 of *Post-War Building Study Number 11*:

> We have reached the unanimous conclusion that a completely new type of 3kW (230 V) socket outlet and fused plug should be adopted as the 'all-purpose' standard.

A member of the study committee, Forbes Jackson, had asked the company in 1943 whether it could design a domestic all-purpose fused plug. Richard Amberton wrote later:[15]

> At that time we did not know that recognised makers of plugs and sockets had already been approached with the same object but without practical results. As soon as it became known that we had succeeded in producing such an article and that it had great merit, we discovered that we had run into a hornet's nest.

The new Dorman and Smith plug was ideally suited to the new arrangement. In the ring main circuit, each plug contained a fuse; Atherton's idea was astonishingly simple – that one of the three pins or prongs of the plug should itself be the fuse (*Plate 17*). Whereas other plugs had to be taken apart to replace the fuse, the Dorman and Smith plug could be repaired simply by unscrewing the pin and replacing it with a new one. The detachable pin was made of hollow tubular aluminium oxide ceramic with a brass cap at the end which was pushed into the socket, and a brass cap and screw thread at the other end to attach it to the plug, with a conical spring to prevent vibration loosening the fuse pin; a fuse wire ran through the centre of the pin. It was inherently safer than a conventional plug because it was extremely difficult to substitute anything for the specified fuse pin, whereas in a conventional plug, dangerous makeshifts such as hairpins could be connected instead of a fuse.

Though the design was simple, the development of the plug was not. The problem lay in the ceramic pin, for it was such a radical change from the conventional metal. The ceramic material was called *Alorite*, since it contained aluminium oxide, and a great deal of time was spent in perfecting it both at the prototype stage and when it had to be formulated in a way suitable for mass production. However, the company had no expertise in chemistry and during the war its funds were scarce, so *Alorite* was developed in a very *ad hoc* way. Atherton had always liked the idea of

[15] R. Amberton, *DS Fused Plug Compared with Plugs to B.S. 1363*, pamphlet published jointly by Dorman and Smith Limited and A. Reyrolle & Co, August 1948.

porcelain and ceramics; possibly he was influenced by the company's early innovations in this field in the late nineteenth century. He talked to a number of porcelain manufacturers but they were reluctant to supply Dorman and Smith because their main customers, the motor car spark plug manufacturers, opposed their supplying outsiders.

No doubt the existing spark plug makers were fearful of Dorman and Smith, which had at the end of 1944 set up a small company, Monarch Spark Plug Company Limited, to exploit Dorman and Smith patents in the manufacture of spark plugs for internal combustion engines. In the event, the electrical side of the group was so busy that Monarch never presented a threat to the established manufacturers. Dorman and Smith had no alternative but to go it alone, and much of the burden fell on Bruce Cooper as joint managing director of DS Plugs. Myles L. Cooper recalls with amusement a visit to his brother at the factory, finding him crouched like a medieval alchemist over a mixing machine full of ceramic ingredients, saying 'let's try a little more of *that*'.

Eventually a compound was formulated which successfully answered criticisms that ceramics would break where metal would not; this was demonstrated by forcing *Alorite* fuse tubes undamaged through quarter-inch mild steel plate. It was, of course, imperative that the fuse ceramic should not break; one publicised failure and the entire innovation would have been discredited in the eyes of the industry. Geoffrey Atherton remembers running up and down the steps of Ordsall station as a schoolboy, dragging a fuse pin plug behind him and vainly trying to break it.

But there were production problems in 1946 because the furnaces used to fire the ceramic were troublesome. The heat was on in other senses too: it was becoming urgent to sort things out successfully to cope with their own flood of orders, and to supply *Alorite* fuses to A. Reyrolle & Company Limited of Newcastle who were negotiating a licence to produce the DS plug. More serious was the developing battle with the rectangular pin plug. Problems remained with the furnaces: the life of the lining was too short, an extra furnace was necessary whenever a furnace was stopped for relining, and the diameter of the *Alorite* tube was irregular and the short run solution of making it oversize and grinding it down to specification size was slow and expensive. However, the company did have a stroke of fortune in recruiting in August 1946 a German chemist called Schoenherr who had worked on porcelains at Bosch, the electrical company which included spark plugs among its products. Schoenherr felt that the existing arrangement, using batch furnaces which had to be loaded, fired and unloaded, was not efficient enough for large scale production. At this time, production targets were becoming ambitious: Reyrolle were thinking of 45,000 plugs and sockets a week, and Dorman and Smith of 20,000, which they would produce in a new factory being built at Salterbeck, near Workington in Cumberland, for which they received Treasury sanction in July 1946. Building licences were strictly controlled and Cumberland was the nearest place to Manchester where Dorman and Smith could obtain a factory with the advantage of low rents charged in development areas. The company agreed with Schoenherr's recommendation that a tunnel furnace, allowing flow production of *Alorite*, would help meet their targets. Eventually the Ministry of Supply located one in Germany, and offered it to Dorman and Smith for about £4,000 to include erection and various extras.

The Salterbeck factory of DS Plugs was formally opened in September 1948 though *Alorite* had been in production there since October 1947, and the last of the equipment from the Pendleton factory had been moved there in April 1948. The factory employed about 240 men, women and girls, and some of them had come with the company from Salford. Bruce Cooper, whose family came of Cumbrian stock, moved to the Lake District and became resident managing director. At the formal opening, Dame Caroline Haslett, a member of the British Electricity Authority Board, praised the product and the plant.[16]

However, it was not all plain sailing. Government restriction of building licences in 1948 bit into demand, and there were contractions and even cancellations of orders throughout the electrical industry. More serious in the long run was the growth

[16] *Cumberland Evening Star*, 7 September 1948.

of support for rectangular pin plugs. There were no technical reasons why rectangular pins were any better than round pins, but many other manufacturers were only too anxious that Dorman and Smith should not sweep the field with their new plug. There was no danger of total monopoly because Dorman and Smith were not big enough and in any event they had licensed Reyrolle to produce the plug, and expressed their willingness to make similar arrangements with other producers.

The board considered various stratagems. It was suggested that the sockets should be sold cheaply, so that local authorities would adopt them for their housing projects, leaving the company to make its profit on the plugs, which were sold more to the householders; but Atherton pointed out that rivals might then undertake licensed plug manufacture, driving profits down and effectively be subsidised by the cheap DS sockets. It was a galling dilemma. Either Dorman and Smith forfeited their reward for the invention, which was so simple and had such a neat patent that it could not be circumvented, or they faced the combined opposition of virtually the whole industry. At first it seemed that they might succeed in the face of the opposition. Forbes Jackson, who had originally introduced the company to the problem of the fused plug, was chief electrical engineer to the London County Council, and specified the DS plug in the Council's housing projects. Other authorities followed suit, but eventually it succumbed to the rectangular pin plug.

However, the opposition preferred the less sophisticated plug with the internal fuse, rather than pay royalties to the firm whose ingenuity had embarrassed them. Consequently, although the fuse pin plug remains in production and has made a valuable contribution to the sales of the group, it has never had the full success that it deserved and most of us still have to take our plugs to pieces to replace the fuses. Even Dorman and Smith were forced to conform and the DS Plugs catalogue now includes rectangular pin plugs, but with the distinctive feature of a hinged pin which enables the fuse to be replaced by swivelling the pin, without using a screwdriver or opening the plug. In summary, the detachable pin plug was a most worthy advance and brought the company a great deal of business, including exports to South Africa, the Middle East, the Far East and Eire; but it failed to sweep the field because of the hostility of opposition from established interests.

While these dramatic developments occurred in DS Plugs, Dorman and Smith continued at Ordsall to make a success of the traditional business in switchgear and lightfittings. The products were continually improved and, as always, orders were won for some of the most distinguished customers, including main switchgear for the new House of Commons and House of Lords, and the electrical equipment for the Bank of England extension. Many more contracts were secured from government-sponsored housing schemes, electricity generating authorities, mines, harbours, railways and shipyards. On the production side, Atherton applied the lessons learned during the war that standardisation paid major dividends: old, slow selling items were dropped and area representatives were appointed to improve the company's market outlets.

The profits reflected the benefits of the new policy and both Dorman and Smith and DS Plugs did extremely well, as the table on page 125 shows.

There was one development during the period that was in one way auspicious but, in another, ill fated. Thomas Atherton visited South Africa early in 1947 and was impressed enough by the prospects there to begin arrangements to set up a subsidiary company. On 17 November 1947 Dorman and Smith (Proprietary) Limited was formed, with a capital of £5,000. It had an office and warehouse in Johannesburg and was intended to sell the products of the English parent company, with the possibility of assembly and even local manufacture in the long run. Although Dorman and Smith had employed overseas agents for decades, this was their first overseas subsidiary; however, this first attempt was later to end in failure, as we shall see in the next chapter.

If we took a snapshot of Dorman and Smith Limited in 1948, we should have a picture of a prosperous and growing company. It had advanced out of all recognition from the company that Thomas Atherton had taken over in 1937, and grown faster than he might have expected; for although the war had posed problems of finance, the pace of activity and the rate of inflation had

TABLE 12 COMPANY PROFITS, 1937–8 TO 1946–7

Year to 31 March	Profits before interest and tax: D & S	DS Plugs	Combined	Profits tax etc[1]	Profits before income tax
1938	6,573		6,573	155	6,418
1939	5,388		5,388	141	5,247
1940	11,009		11,009	2,583	8,426
1941	12,023	3,139[2]	15,162	5,941	9,221
1942	13,412	12,865	26,277	15,916	10,361
1943	19,640	11,640	31,280	18,371	12,909
1944	33,346	20,639	53,985	39,945	14,040
1945	35,895	9,976	45,871	30,381	15,490
1946	24,036	3,041	27,077	11,057	16,020
1947	31,342	25,310	56,652	20,400	36,252

Notes: 1. Comprising national defence contribution, profits tax and excess profits tax.
2. 26 August 1940 to 31 March 1941.

reduced the mortgage debenture from being a pressing problem in 1937 to a relatively trivial issue when the final amount outstanding was repaid in 1948. The alliance between Atherton and Myles Cooper in 1937, with the former providing entrepreneurship and the latter capital, had many possible seeds of discord. Atherton was young, driving and ambitious, while the Coopers were a family 'with generations of experience as minority shareholders, and the fullest knowledge how to protect their interests' as one member of the family put it. Yet while there was debate and even disagreement over various issues from time to time, the sum effect was harmony. The son of Myles Cooper, Bruce L. Cooper, entered the company and had shown himself an excellent executive director, and his brother, Myles L. Cooper, had given useful service as an outside director and was one of the initiators of the idea of a holding company.

Earlier in this chapter, we saw that Atherton in 1937 was interested in eventually gaining the company a stock exchange quotation, as a means to capital gain. That was at a time when he was financially restricted, but by 1948 the success of the company had effectively freed him. His decision to offer preference shares to the public was mainly influenced by the need to provide an outlet for shares, for though he was not yet forty-nine years old, it was only prudent to take precautions against premature death. Death duties were heavy, and anybody with the vast bulk of his assets in a private company was at a potential disadvantage.

Moreover, though the company had done well, the war had limited the extent to which Atherton and Bruce Cooper could earn large salaries and dividends. In 1947 both were favourably impressed by a suggestion that a holding company be established by a placement of shares quoted on the Manchester stock exchange, since it could be arranged to yield cash to them after many years of hard work and restricted reward. The story of the holding company and its subsequent growth and success are the subject of the next chapter.

CHAPTER FIVE

The Public Company: Growth and Maturity

The new company was registered on 9 April 1948 as Dorman Smith Holdings Limited. For the first ten days of its life it was a private company, but was then converted into a public company. A holding company is one which holds its main effective assets in other companies, which are usually subsidiaries; though it may have offices, property, cash and even operating divisions of its own. This form of organisation spread in the nineteen twenties and thirties, but according to the *Investors Chronicle*:[1]

> What started as a measure of obscurantism soon proved to have real advantage as a type of organisation. To vest a well-defined part of a business in a separate company simplifies management, costing and administration and may have advantages in customer relationships as well as in legal matters.

Undoubtedly, the new form of organisation did prove advantageous in the years to come. The original subsidiaries were four: Dorman and Smith Limited, DS Plugs Limited, Dorman and Smith (Pty) Limited in Johannesburg, and Monarch Spark Plug Company Limited. The shares that the Atherton and Cooper families held in the companies were sold to the holding company, which in return granted shares to the Athertons and Coopers. None of the ordinary shares were to be sold, but it was intended that up to 120,000 of the total of 175,000 £1 six per cent preference shares in the new company would be sold after a placement

[1] Investors Chronicle, *Beginners Please*, Eyre and Spottiswoode, London, 1960, p. 155.

by the company's brokers, Dimmock and Cowtan. They were expected to command a premium by fetching at least 22s 6d, for although they were not to be offered for sale to the general public, they were an attractive low risk investment at a time when bank rate was only two per cent. The Prudential Assurance Company became the largest outside shareholder. The advantage to the company of a placing was that it was cheaper than an open issue for sale.

The new arrangement meant that the preference shares could be dealt in on the Manchester stock exchange. This made them easier to sell should Thomas Atherton or Bruce Cooper wish to do so in the future, but it also realised cash for them, for between them they subscribed for about 67,000 of the preference shares in the new company at 22s 6d, and so raised over £75,000 for it, but were also allotted extra preference shares in exchange for their previous holdings in Dorman and Smith Limited, DS Plugs and Monarch. Some of these extra shares and their newly bought ones were then placed by the sharebrokers, providing immediate cash for Atherton and Cooper, and more than covering the cost of subscribing to their 67,000 preference shares. Thus in a neat package deal, Atherton and Cooper raised fresh capital for the company, capitalised some of their own shareholding and retained control of the organisation. The new preference shares, which were cumulative and redeemable, carried no voting rights of course, and Thomas Atherton with his wife and children held 61½ per cent of the ordinary shares. Bruce Cooper held almost 28 per cent of the ordinary shares, and relatively small holdings were allotted to other members of the Cooper family and to Amberton, Bickerton, Dale, Lund, Nesbitt and Smalley.

In their first year as subsidiaries of the holding company, both Dorman and Smith Limited and DS Plugs did well and comfortably exceeded the profit forecast at the time of the flotation. It is perhaps ironic that Thomas Atherton, who had for so long been restricted in his income from the business, was now entitled in the year ending 31 March 1949 to a total pre-tax income of over £9,000 from his salary, commission and dividends from the enterprise, although this level was not sustained in the following year.

PLATE 22. The Atherton Works at Preston, purchased in 1958

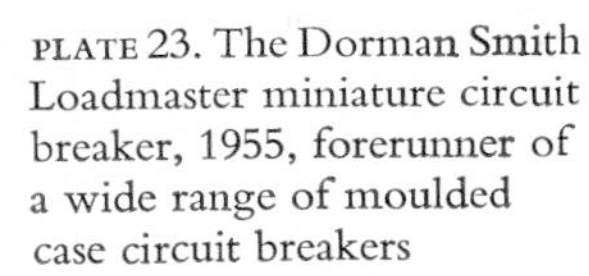

PLATE 23. The Dorman Smith Loadmaster miniature circuit breaker, 1955, forerunner of a wide range of moulded case circuit breakers

PLATE 24. Processing *Alorite* ceramic fuse barrels at the Workington factory in Cumberland

A bargain acquisition: British Klockner Switchgear

In August 1948 the Board of Trade invited tenders for British Klockner Switchgear Limited, a manufacturer of electrical control devices, notably motor control gear, with a small factory in Chertsey, Surrey. It had a short but interesting history, having been established as a subsidiary of F. Klockner, KG of Cologne. It was formed with the intention of acting as a sales, stockists and service depot for the German parent company which had previously employed an agent with the intention of manufacturing and assembling certain of the parent company's products as finances and conditions permitted. The second world war severed the German connection but the company continued to manufacture using local sources of supply. The manager, Gilbert A. Standen, wished to preserve the company that he had created but to do so he had to walk a financial tightrope: if he succeeded in making the business noticeably profitable, the British authorities might seize it, but if he was too lax there was a real risk of bankruptcy. However, he was not only able enough to survive the war but also to continue as managing director after the takeover by the new company, and within four years became a director of Dorman Smith Holdings.

On the face of it, British Klockner Switchgear was not particularly impressive. The factory was only about one-twelfth the size of the works at Ordsall, with a total employment of twenty-nine and net tangible assets of £13,500. However, it had an excellent profit record, culminating in a pre-tax profit of nearly £12,000 in 1947. Atherton persuaded his fellow directors that the company had excellent prospects and would be a good buy. Although it was small by their standards, the product line would be a useful adjunct to the Dorman Smith range, as Klockner were specialists in motor control gear, contactors and ancillary switches for the control of machine tools, process plant and lifts (passenger and goods). He persuaded the board that they should be prepared to pay up to £25,000 but in the event they had to pay only £17,500. The extent of the bargain becomes clearer when one examines the subsequent profit record: in 1949–50, the first full year of operation as a Dorman Smith subsidiary, British Klockner

Switchgear profits before tax were higher than the purchase price and in two years it had virtually paid for itself. In 1951–2 it had an extremely profitable year and earned almost as much profit as Dorman and Smith Limited and DS Plugs combined, for the profits of those two subsidiaries had declined as profit margins came under pressure. To sharpen the comparison, we can examine profit rates in the three subsidiaries in 1951–2:

TABLE 13 PROFITABILITY OF SUBSIDIARIES, 1951–2

Subsidiary	Pre-tax profit as a percentage of:	
	Sales	Capital employed
Dorman and Smith	6.4	11.6
DS Plugs	5.8	10.2
British Klockner Switchgear	31.6	99.1

While this was an exceptional year, the profitability of British Klockner Switchgear remained well above the group average throughout the nineteen fifties and the first half of the sixties and made an invaluable contribution to the holding company. Major new buildings were erected at Chertsey between 1954 and 1958 and the company grew very considerably. Without any doubt, Atherton's purchase in 1948 was a most remarkably successful and timely addition to the holding company, the more so when seen against the background of the problems that emerged in the first half of the nineteen fifties.

Problems in the early fifties

The group faced a wide range of problems in the early years of the nineteen fifties. Practically the whole of the time that Atherton had been in control since 1937, it had faced a seller's market: during the period of rearmament and the war, the burning problem had been how to produce goods, not how to sell them, and though the company had to readjust to peace time markets, it had done so successfully, for it emerged from the war with good productive facilities and a novel product, the fused plug. But by the end of the forties, trading conditions changed. Competition increased in many of the standard lines produced at the Ordsal Works, such as handlamps and *Diolux* lightfittings. The minutes of the directors' meetings at the time show continued anxiety at the sluggishness of sales in standard lines.

The position would have been transformed had the company's patent plug with the fuse in the pin been widely adopted, but the combined opposition of all the major producers in the industry, who favoured the unpatented flat pin plug, overwhelmed Dorman and Smith. So although by October 1948 about a million DS sockets had been sold, the chairman noted disconsolately in a memorandum to the directors of DS Plugs:

> It looks to me as if our trade in plugs and sockets is dying . . . I think the reason that we have not gone into the manufacture of the flat pin plug is because we felt we would lose a certain amount of 'face'. I really do not think we should worry about any loss of face providing we are making money, and therefore we have to see things as they are . . .
>
> I am not suggesting that we should drop what we are now doing, and immediately go on to production of these new plugs. I feel that the programme we have commenced should be regarded as sacred, and that production of an adaptor and a tumbler switch and a switched plug should proceed with speed, but that in addition we should regard the production of these other designs of plugs to be undertaken as a matter of urgency.

In some senses, this decision came a little late, for as recently as April Atherton had decided against jumping on the flat pin bandwagon straight away because his sales representatives had still been optimistic about the Dorman Smith plug. It was a galling and disappointing decision for him to have to make, but once the British Standards Institute had come down in favour of the flat pin fused plug, as BS 1363, there was practically no choice. However, the interval between concentrating on the DS patent plug and the flat pin was most uncomfortable. By January 1949 the Workington factory of DS Plugs was almost at a standstill and forty-five people had to be given notice. Sales of the subsidiary were more than halved between 1948–9 and 1949–50, and the profit margin on this reduced turnover was almost halved. It was more than a decade before the sales of 1948–9 were matched.

Although, as we have seen, British Klockner provided some welcome relief in an otherwise gloomy outlook, it was essential to get the main body of the business on to a firm footing of growth again. This was attempted in two main ways; by putting an increased emphasis on marketing and by improving the range of products that the company sold, including some new ones.

Both DS Plugs and Dorman and Smith Limited introduced a number of measures to try and improve sales. To give potential clients a better idea of the product range, DS Plugs fitted out a van as a mobile showroom equipped with facilities for short circuit testing.[2] Supervised by a competent technical demonstrator, it called at factories, shipyards, mines, technical colleges and trade shows and fairs and generated considerable interest in the company's products. Dorman and Smith Limited cut prices on their *Diolux* range of lightfittings to meet competition. At Amberton's suggestion, they published a booklet, *The Contractor's Friend*, which showed their products, along with electrical tables and regulations. They extended the discounts given to the British Electricity Authority, reorganised the London sales arrangements by taking over Amberton's agency and appointing Robert Speirs as London sales manager responsible for London and eighteen southern counties, and rented teleprinter facilities to improve communications with Salford. More generous discounts were given to wholesalers. But standard lines sales still flagged, notwithstanding the introduction in late 1952 of a cheap line of lightfittings. In 1953 four speciality salesmen were appointed and given intensive training before being put in the field to boost sales of lightfittings; however within less than a year three of the four had been dismissed and a further blow occurred in

[2] See *The Electrical Times*, Vol. 116, 18 August 1949, p. 227.

1953 when the London County Council announced that it was giving up the DS plug and specifying flat pin plugs in its housing developments. In May and June of 1953 Dorman and Smith Limited sustained losses of about £7,000; part of these was due to the acquisition of Oldfield Engineering Company Limited, but the trend was serious enough to prompt reductions in staff and overheads in the hope of saving £1,500 a month. The following profit figures show how the various major subsidiaries and the parent company fared:

TABLE 14 GROUP PROFITS, 1948–9 TO 1954–5

	Pre-tax profits (£) of:			
Year ending 31 March	Dorman and Smith Limited	DS Plugs	British Klockner	Dorman Smith Holdings
1949	24,545	35,823	—	64,948
1950	23,308	8,072	17,881	47,035
1951	14,432	16,724	20,581	51,595
1952	25,402	8,248	33,042	63,296
1953	14,093	18,981	27,713	57,681
1954	1,586	28,618	21,907	43,219
1955	10,907	26,752	39,213	76,505

The financial year 1953–4 was clearly a low ebb in the fortunes of the group. The figures recorded must be viewed in the light of the prospectus issued on the formation of the holding company in 1948, which mentioned profit from the subsidiaries exceeding £60,000 in 1947–8. Yet even with the inclusion of British Klockner, this figure could not be matched six years later. At this point it is relevant to consider one of the contributory factors to the decline in profits.

Failure in South Africa

As mentioned in Chapter Four, Dorman and Smith (Proprietary) Limited was set up in Johannesburg in November 1947, with an office and warehouse to sell the products of Dorman and Smith and DS Plugs and possibly to assemble and manufacture if this became economic. The issued capital was £5,000 of which £4,998 was subscribed by Dorman and Smith Limited and DS Plugs. However, problems rapidly arose. The South African government imposed import licences on all imports entering the country after 13 June 1949, which severely limited the scope for expansion of business. Usually such quotas restrict imports *pro rata* to past imports, and tend to hinder newer companies, such as Dorman and Smith (Proprietary) Limited, though the shortages that follow often lead to an element of monopolistic profit. But before this situation could arise, the subsidiary incurred a loss of £3,406 in 1948–9 to the surprise and concern of the directors in England. In January 1950 a payment of £4,500 was made to the South African company, though Atherton and his fellow directors in England remained uneasy about it and hoped that the management in Johannesburg would run down their stock so as to keep total commitments below £20,000 – for not only was there direct investment of £9,000 but extensive debts for goods supplied from England. In 1950–51 the South African company lost £2,000 and towards the end of 1951 invoices from England were not being paid. In November, Amberton sailed for South Africa to investigate matters first hand.

He was not very impressed by what he found but Atherton was even more sceptical, saying that the board had lost confidence in the South African management because of its failure to send figures and its failure to carry out its other promises. He was anxious to cut the parent company's losses and run the South African interest on an indent basis. In the year 1951–2 a loss of £3,664 was sustained, giving a cumulative loss approaching £10,000. Total indebtedness to the parent company, including the original investment and monies owing, approached £40,000. While he was in Johannesburg, Amberton had discussed a re-organisation scheme which would require £40,000 but this was vetoed when he returned to England. A proposal from Johannesburg for association with another South African company was also rejected because it seemed unlikely to repay the debts owed to the parent company, which were now so large that, had they been owed to an outside company, liquidation would almost certainly have followed. However, as the largest creditor, Dorman Smith Holdings hoped to salvage something by selling the stock held in Johannesburg. The manager there, who was not in good health, gave up his full time interest at the end of 1952 and his assistant succeeded him.

Amberton sent explicit instructions to Johannesburg outlining the savings that would be necessary to make the business viable; they were to be effected by halving the staff, moving premises and selling certain buildings; regular monthly statements were to be sent to England and the first priority was the repayment of the bank overdraft. The new manager was given a year's grace to bring about the transformation, but sales were not high enough and he resigned after eighteen months. Denis Kinnell, group export manager and later a director of Dorman Smith Switchgear Limited, visited South Africa and arranged the sale of the company in early 1955 for just under £4,500. The parent company wanted the name of the company to be changed so as to exclude 'Dorman and Smith' but to do so would have meant forfeiting valuable tax losses, so that eventually it became Dorman Electrical Sales Company (Proprietary) Limited. The continuance of the name was a contributory factor in Dorman Smith Holdings renewing their financial stake under happier circumstances in 1964.

The South African experience of the first half of the fifties was almost traumatic for the English management. Fifteen years later it could still cause head shaking among those who remembered it. Thomas Atherton admitted that he had made a basic mistake over the venture, and Robert Dale said it was the worst problem that they faced during his thirty years as company secretary. However, a lesson was learned and the company never again allowed itself to be lured into such an exposed position.

New product developments

It would, however, be misleading to suggest that all was unrelieved gloom in 1954, for within the company important developments were taking place which paved the way for future successes. The development of the fused plug had not brought as much commercial success as the company had hoped, but it gave a subtle psychological boost; the company had beaten the giants in developing the plug and only been restrained from total success by the weight of numbers of the opposition. As the market for fused pins had failed to grow as the company had hoped, other outlets were sought for the *Alorite* ceramic. In this, Bruce Cooper at Workington made a major contribution to the group by successfully breaking into the market for high rupturing capacity, or HRC, fuses. The field had been dominated by English Electric, so the breakthrough added further to the confidence of Dorman Smith Holdings Limited that they could take on practically all comers.

In the fifties, a number of designs were produced which yielded increasingly important business to the group. The first of these was the *Safeset* miniature circuit breaker, designed in 1950 by H. W. Wolff.[3] Traditionally, British electricians had preferred fuses to circuit breakers, though the latter were popular in the United States and Continental Europe. The circuit breaker has the advantage for the user that he need only press a button to restore the circuit after a malfunction, whereas mending a fuse means fiddling with wire and screwdriver, or at best, replacing a cartridge fuse. However British interest in miniature circuit breakers[4] grew in the post-war years, and *Safeset* was the first one to be designed, developed and manufactured by a British company. Sales grew rapidly and reached record levels in December 1952.

Unfortunately the company became involved in a long legal battle with the American Heinemann Company, which had filed a patent application on miniature circuit breakers. Dorman Smith developed its breaker independently but the patent eventually granted to Heinemann was much stronger than the original

[3] The new circuit breaker was reported in *The Electrical Times*, Vol. 120, 6 December 1951, p. 1037.

[4] There was much opposition to them among sections of the trade; for a contemporary article supporting them see T. C. Gilbert's 'We lag behind in Efficient Electric Circuit Protection', *The Architect and Building News*, 7 October 1954.

application and Dorman Smith were held, after appeal, to have infringed it. Heinemann agreed to drop an action in 1953 on payment of £750 and again in 1956 on payment of £1,500. The controversy absorbed a good deal of the directors' attention and the company was perhaps unlucky; but it never relaxed its intention to push ahead with circuit breakers and did all it could to popularise them.[5] The benefits of this determination are still accruing today.

Another important development which appeared in 1951 was the TS switch, named after its designer, Thomas Smalley, who completed the prototype shortly before he retired as chief designer. Usually a handle was put on the side of the box. Used singly, this was unimportant but when boxes are ranged alongside each other in a multiple installation, the space taken by the handles is waste, and in a large installation the waste could be considerable. Smalley's contribution was to put the handle on the front of the box; it was not a complete novelty, as similar designs already existed in the United States, but Dorman and Smith were in the forefront of British manufacturers, and the TS range was refined and extended by Smalley's successors and remains in production today.

In 1954 the chief designer, W. A. (Arthur) Cockroft, developed the Dorman System of switchboard construction. Traditionally switches had used cast iron boxes, or 'ironclad' as it was commonly termed. The Dorman System joined the trend away from castings to pressed steel sheet, which is much easier and cheaper to make; but the distinguishing feature was its modular characteristic, so that basic units could be built up to virtually any size that the customer required, rather like a Meccano or Lego set. This standardisation-with-flexibility appealed both to the customer and the production department, since it kept costs down. Two years later, Cockroft developed the concept further with the *Cubicon* system of cubicle switchboards, again using a modular approach. Customers could buy the parts and assemble their own cubicle switchboards to satisfy their specific requirements; it was especially

[5] For example, Mr Wolff gave a public lecture demonstration in July 1953, *Electrical Review*, Vol. 154, 10 July 1953, p. 111.

attractive to contractors who could employ their own labour on assembly. The new systems had the advantage of rapid assembly, and a special demonstration of this was given at Prestwick Airport in 1955 when an extensive switchboard was installed for the US Air Force.[6]

Development also continued with the miniature circuit breaker, and in 1955 the *Loadmaster* was introduced. T. G. F. (Geoffrey) Atherton, the chairman's elder son, was mainly responsible for it, assisted by W. E. Minoprio. It was more compact than the earlier *Safeset* at a time when miniaturisation was becoming increasingly important, and remains an extremely important part of the company's business today. In a sense, Geoffrey Atherton cut his business teeth on miniature circuit breakers; he and Wolff lectured on the virtues of the breakers to meetings of the Institution of Electrical Engineers up and down the country; he handled the long and delicate negotiations over the Heinemann patent; and he developed the company's own breakers, and later extended the range to carry higher current loads.

Thus by the mid-fifties, the company had a good range of switchgear products and was set to take advantage of the growing trend to use circuit breakers instead of fuses. All the new products were smaller for the same functional capacity than their predecessors, and this not only meant savings in space but savings in sheet steel and other materials such as the copper used in busbars within the switchboard.

However, new products are only half the key to success and, as the recent Rolls-Royce collapse shows, it is vital to produce them efficiently. As we saw earlier, Thomas Atherton was pruning staff and overheads in 1953 to try and limit rising costs. However, unlike his predecessor as chairman, H. G. Baggs, Atherton was prepared to take good advice and to pay for it. In April 1954, Associated Industrial Consultants offered to conduct a seventy-week study of Dorman and Smith Limited for £8,800, as a result of which they expected to be able to effect savings of £12,000 a year. They proved extremely valuable and, in the words of Geoffrey Atherton, 'they taught us how to run a factory'; but

[6] *The Electrical Times*, Vol. 128, 24 November 1955, p. 816.

this was not a platitude, for competitors had also employed consultants yet still succumbed to competition. Once their worth was established, consultants were employed widely in the organisation – on advertising, sales, budgetary control, stock control, costing methods and manufacturing methods. Eventually, consultants were also employed on marketing, though the management was rather apprehensive since they felt that in this area it was difficult to keep a check of the consultants' progress, whereas, in the factory, management could see quickly if anything was seriously amiss with the consultants' policies. However, in the nineteen sixties, according to Geoffrey Atherton:

> quite simply, they taught us how to sell. They set up a scheme so that the representative knew how he was doing, and he also knew that we knew how he was doing.

Other members of the company say that the consultants won their full respect and co-operation, and that this helped promote successful operations.

Thus as the decade progressed, the company strengthened its product line and improved its management expertise. Since then, almost uninterruptedly, the profits of the group have risen steadily.

Acquisitions and new companies

For the past two decades, Dorman Smith Holdings has been, like many other successful companies, the subject of many invitations to buy other smaller businesses. The story of all these – the motives and negotiations – would be a sizeable study in itself, so that here we can only chronicle the ones that eventually reached fruition.

At the end of 1949, the directors of the holding company were considering alternative sales outlets for *Alorite*, the ceramic element of the fuse pin in the DS plug. Capacity was necessary to make the most of the tunnel kiln, yet sales of the plug had been adversely affected by opposition companies, as we saw earlier. Various possibilities were examined, including cartridge fuses, spark plugs, *Alorite*-tipped tools, guide eyes for textile machines, insulating sleeving for electrical switchgear and as a grinding medium. Some, such as spark plugs, fell by the wayside, but it was felt that the product stood a better chance if it were marketed by a company with no apparent connection with the electrical industry and in 1950 the Monarch Spark Plug Company was renamed Alorite Limited. Business built up gradually through the fifties and produced a good return on capital, though it never constituted more than a small fraction of the total sales of the group. *Alorite* is currently used for Dorman Smith fuse barrels.

Problems in getting supplies of iron castings from the Midlands prompted an acquisition in 1951, when the company bought John Booth Foundries Limited of Derby Street, Preston. It was significant in more ways than one, for not only did it meet the immediate needs for castings but it played a critical role in the eventual move of the bulk of the group to Preston. The purchase price was £11,000, but its early years were not profitable. In 1954 Clarence Bickerton, who had been with Dorman Smith for twenty-three years, was appointed managing director of John Booth and conscientiously proceeded to build it up into a very profitable small subsidiary and win the universal admiration of his fellow executives. By casting non-ferrous metals as well as iron, and by doing sheet metal pressing, it prospered in a period when other iron foundries were driven out of business, and today most of its turnover comes from outside the group.

While the Ordsal Works had been described as the most complete of its type in the country sixty years before, during the nineteen fifties its shortcomings were all too apparent. It had three storeys and a basement, when modern factory practice specified one storey layouts; and the Hampson Street works were across the road. The company had bought some neighbouring cottages in the eventual hope of rebuilding, but it could not get permission to demolish them. As early as 1943, architects had been consulted about the possibility of a completely rebuilt factory on the site of the Ordsal Works, but it was felt that the disruption of production would have been too severe.

In March 1953 Thomas Atherton was told by Mr Davy, tht managing director of Oldfield Engineering Company Limited, that his factory in Ordsall was up for sale. Oldfield specialised in making coils and transformers, reconditioning electric motors and rewinding, and was of no real interest to Dorman Smith as a going concern, but its premises were only 400 yards away. They included a building of five thousand square feet with a five-ton gantry crane and ample clearance, whereas Dorman and Smith were having increasing difficulty in assembling their bigger switchboards and loading them on board lorries. Their multi-storey factory had low ceilings and they had rented a disused chapel from British Railways but its floors were weak, the buildings were badly maintained and the rent was about to be doubled. Atherton offered Davy £12,500 for the business, and the new premises gave Dorman and Smith a valuable breathing space, though within a few years the group was looking at other sites to build a new factory. Some of the Oldfield plant was sold, but realised less than was hoped, so it was decided to keep the rest as a going concern with Eric J. Atherton, the chairman's younger son, and Mr G. Hatton, the previous works manager, as joint managers. However, the business lay outside the group's line of experience and the building and assets were eventually sold in 1957, but Oldfield Engineering was reconstituted as Dorman and Smith (Domestic) Limited and began operations in Preston in 1959.

After their South African experience, the directors of Dorman Smith were not enthusiastic about foreign manufacturing ventures, but the export business of the company inevitably raised the question from time to time. For some time the group had considered collaborating with Ratansi Morarji Limited in India. In June 1957 British Klockner formed a joint venture with the Morarji family in Bombay. It was called Hindustan Klockner Switchgear Limited, with an issued capital of five million rupees (or about £37,500) of which British Klockner held forty-nine per cent; however, there was a major difference from the South African arrangement, since British Klockner was granted its share without actually subscribing a penny but simply in return for guaranteeing technical advice to the Indian company. In this instance, Thomas Atherton was not the driving force. He later recorded:

> I simply sat down and did nothing. The know-how was supplied by the Chertsey end. It all came out very well eventually; though the return on capital wasn't high, it hadn't really cost us anything.

In January 1961 the Preston side of the business took a similar forty-nine per cent interest in Morarji Dorman Smith Limited of Bombay, which had an issued capital twice the size of Hindustan Klockner. The new company was to manufacture the *Loadmaster* miniature circuit breaker and was free to make all the British lines of Dorman and Smith Limited and DS Plugs. Dorman Smith Holdings were not to participate beyond £10,000 and free technical information formed the other part of the consideration. Morarji sent a man to Preston to train in manufacturing circuit breakers. Since import was virtually forbidden into India, Morarji hoped to gain a virtual monopoly of the market. Although it took a couple of years to earn profits and labour productivity was very low by British standards, Morarji Dorman Smith, like Hindustan Klockner, later paid dividends to Dorman Smith Holdings and also provided a market for British component parts which would have been closed if there had been no manufacturing interest in India.

The company's products, especially miniature circuit breakers, did well in export markets as the decade progressed, particularly in Australia and New Zealand. A Brazilian company acquired a licence to produce Dorman and Smith products and in 1959, an Israeli company applied to undertake licensed manufacture but the board reluctantly declined because after consultations with the Board of Trade they judged that to grant a licence would provoke a boycott of Dorman Smith products by the Arab states, which also figured among their export markets.

THE YEARS OF GROWTH, 1955–71

After the low point of 1954, the company seemed to find second wind. The reasons for this are not clear cut and many contributory factors can be seen, not least of which is the potential which, as we have seen, already existed. In 1955 Charles Nesbitt retired as sales manager and director and Denis Kinnell, who had been group export manager, succeeded him as sales manager. He quickly introduced a revised system of commission for sales representatives and proved his ability in the job. In 1958 he was appointed to the board of Dorman and Smith Limited.

The second generation of Athertons

Thomas Atherton's two sons, Geoffrey and Eric, played a growing part in the business, as their father had hoped they would. Thomas Atherton recalls with evident pleasure and amusement how Geoffrey as a small boy before the second world war, had reacted to the promenade illuminations at Blackpool by exclaiming: 'There are hundreds of thousands of millions of volts here, Dad!' Of the two sons, Geoffrey proved the more interested in the engineering side, though both worked in the factory during school holidays. Eric Atherton became a Fellow of the Chartered Institute of Secretaries after they left Giggleswick School, while Geoffrey worked for a year at the Ferranti factory in Hollinwood, living as a lodger in a council house and earning £2 10s 0d a week. He then went up to Clare College, Cambridge, where he divided his energies between rowing and the Mechanical Sciences Tripos from which he graduated with honours in 1951. No less influential than his time at university was the year he spent in the United States. His father knew L. W. Cole, the president of Federal Pacific Electric, a major electrical manufacturer, and arranged for Geoffrey to meet him when Cole visited London. Thomas Atherton ranks this with the later move to Preston as one of the best things he ever did for Dorman and Smith. Despite the level at which the arrangements were made, there was no red carpet for the young Atherton: he took his union card and worked at a bench on the shop floor. In his own words, he benefited enormously and came to appreciate the importance of production engineering and the potential of the mass market which the United States exemplified. This applied particularly in the case ot the *Stablok* circuit breaker invented by the president's son, Tom Cole: it was a phenomenal commercial success, cutting costs by three-quarters and taking Federal from the minor to the major league. Against this background, his response to an approach by his father was not surprising: Thomas Atherton said in 1970:

> At the end of his year at Federal, I rang him up and asked him if he would take charge of the circuit breaker department if he came over to this country; he said he would. He came over, took charge of miniature breakers, designed his own and sold it with the result that it's selling over a million pounds a year now.

Geoffrey Atherton's time spent in America has had other benefits for the group, for his later return visits brought in new ideas, and were directly responsible for the establishment of a new and successful subsidiary, as we shall see presently.

In October 1955 Geoffrey Atherton was appointed to the boards of Dorman and Smith Limited, DS Plugs and British Klockner Switchgear. Three months later Richard Amberton resigned his directorship with the group at the age of seventy-three. Thomas Atherton lost a long-standing associate who had done excellent work for Dorman and Smith for thirty years; but the advance of his son, Geoffrey, was the beginning of still greater things.

The move to Preston

As the group's financial performance improved after 1953–4, its cash balances accumulated and the confidence of the directors grew. Liquid assets of the holding company were £143,000 on 31 March 1958 compared with only £46,000 four years earlier. Numerous proposals to bid for smaller electrical companies were examined and turned down, and the idea of an entirely new factory became increasingly attractive: the Salford premises were old, fragmented and increasingly unsuitable for a growing company. In March 1957 Thomas Atherton asked for bores to be sunk on a site at Southport to see if it was suitable for a factory. The issues considered by the board included the likelihood of getting permission to extend the factory in the future; the availability and suitability of local labour; and transport facilities.

PLATE 25. Extruding an *Alorite* section. These sections are later cut, fired and incorporated into high rupturing capacity fuses

PLATE 26. The Dorman Smith transistorised TrafiLAMP, now in use throughout Britain and with extensive export markets

Probably no less influential in the initial choice was the fact that several of the directors within the group lived in Southport and had done so since the days of C. M. Dorman and R. A. Smith – attracted not only by the social standing of the resort but also its low rates and subsidised train service to Manchester. One site turned out to be waterlogged, and others suggested by the Borough Architect usually required piling. Early estimates of piling costs were £10,000 to £12,000 but they suddenly escalated to £35,000; Southport was out of the running. The Board of Trade suggested Skelmersdale, between Southport and Wigan, which was to become a 'new town' in 1962.

However, lengthy deliberations were cut short by the opportunity to buy an existing factory. Clarence Bickerton had been looking for bigger premises for John Booth Foundries, the Preston subsidiary of the group, and mentioned to Thomas Atherton that he had looked at Embroidery Mill, a modern single storey factory on Blackpool Road; it was being offered for sale by a subsidiary of Cyril Lord Carpets but it was really too big for John Booth. Atherton immediately went to see it and urgently commended it to his directors. The asking price was £100,000 but Dorman Smith's bid of £78,500 was accepted in April 1958. Compared with the 63,000 square feet of the Ordsal Works its 125,000 square feet on one level seemed vast. The secretary, Robert Dale, found it enormous and thought that they would never fill it, but he was glad to quit Salford; apart from the operational disadvantages of the Ordsal Works, it was in a slum area and rates were high and he had a private fear that the old works could be a fire trap, with its three old wooden floors soaked in oil from three generations of work in the machine shops.

The changeover from Ordsall to Preston posed many problems. Machinery would have to be moved; likewise key personnel with all the problems of rehousing; local labour would have to be trained; modifications were necessary at the new factory since it had no office accommodation; and the old premises in Salford had to be sold. Rates and heating costs for the new works would also be higher. The direct cost of the move amounted to about £39,000 including gratuities to employees who were leaving the company, and allowances to those who were moving to Preston, plant removal costs and legal expenses, and £27,000 to be spent on alterations at the new factory. However, consultants estimated that there would be annual cost savings of £15,800 at the new location, though it took about a year to get labour productivity up to the old levels.

Preston Corporation was very helpful and provided about twenty houses for Dorman Smith employees to rent. The company gained the co-operation of an assurance company to finance employees who wanted to buy houses. About a dozen of the senior staff and almost all the foremen moved to Preston. Many of them were virtually irreplaceable. The works manager, Max Hoggett, had joined the company as a nineteen year old draughtsman in 1937 and now brought with him to Preston the loyalty and enthusiasm which had proved so valuable to the company. Stanley Royds, whose career had been interrupted by war service, continued as contract sales manager – doing as one director put it, 'an absolutely superb job'; as did W. E. (Ted) Williams, the company's buyer.

Loyalty ran so high that several employees on the brink of retirement elected to carry on at Preston: George Wood, who had joined the company in 1915 and can still recall C. M. Dorman and R. A. Smith vividly, came with the company having served as buyer at Ordsall for many years; and Stanley Johnson, a foreman, commuted to Preston from 1958 until he retired after fifty years service in 1965.

A specialist transport company moved all the machinery during the works annual holiday in August 1958. For about six weeks beforehand a busload of forty Preston girls was taken daily to Ordsall to learn how to assemble the company's products; and when the new factory opened, there was no shortage of recruits – indeed, women were queuing up for jobs.[7] One of them, unknowing, asked Geoffrey Atherton how he had got a job there, and was told – 'well, it's rather a long story!'

Inevitably, the cost of the move fell upon one year in the group's accounts, 1958–9, though the benefits are still being felt

[7] *The Electrical Review*, 26 December 1958, p. 1198, noted that the factory had provided work in an area where there was much unemployment.

today. This can be seen in the table below:

TABLE 15

FINANCIAL RESULTS, DORMAN SMITH HOLDINGS LIMITED, 1954–5 TO 1970–71

Year to 31 March	Profit before tax (£)	Capital employed[1] (£)	Rate of return before tax (per cent)
1955	76,505	445,432	17.0
1956	103,118	484,502	21.3
1957	131,351	525,337	25.0
1958	139,944	570,990	24.5
1959	100,531	595,515	16.9
1960	184,718	647,465	28.5
1961	234,211	909,841[2]	25.7
1962	300,915	928,916	32.4
1963	301,385	1,001,132	30.1
1964	301,572	1,067,795	28.3
1965	355,810	1,142,166	31.2
1966	302,185	1,192,588	25.3
1967	311,779	1,244,274	25.1
1968	376,655	1,320,357	28.5
1969	441,043	1,410,970	31.2
1970	560,150	1,522,958	36.8
1971	834,821	1,724,977	48.4

Notes: 1. Defined as issued capital plus reserves.
2. Assets revalued.

The new factory was appropriately renamed the Atherton Works. The method of its selection makes an interesting contrast with the precepts of economic theory, which suggest[8] that

[8] See, for example, David M. Smith, *Industrial Location*, John Wiley, London, 1971.

firms will seek an optimal location where revenue exceeds costs by the greatest possible margin. Other theorists suggest that businessmen 'satisfice', that is, they seek a satisfactory rather than a uniquely maximising solution to their problems.[9] The choice of the Preston works comes much closer to the second principle than the first, for by no means all potential locations had been studied but it was quickly apparent to Thomas Atherton that the Preston factory was a good prospect. In these circumstances, a quick decision was required, and made; moreover, precise calculations about profitability of the move could not be made because they depended on the prices realised for the three properties to be sold in Ordsall. The Oldfield Works and Hampson Street were sold reasonably quickly for more than their book value, but the Middlewood Street works remained unsold until 1962, when it brought a very low price; overall, the three Ordsall works fetched only £34,300 against their 1958 book value of £41,500.

Progress in the 1960s

The profit figures shown in the last table are a remarkable record, particularly in the latest years when they are virtually without parallel in the long history of Dorman and Smith. Much the biggest source of the increase was from the 'oldest' member of the holding company, Dorman and Smith Limited.

Table 16 only shows relative rates of growth, but not the absolute importance of the subsidiaries concerned. In terms of profit earned in 1967–8, Dorman and Smith Limited was two and a half times as important as the other four subsidiaries combined. Rapid rates of growth such as the group has achieved in recent years can only occur if one or more of a number of conditions exist. There may be an increase in the demand for the basic products of the group; they may match or perhaps beat their competitors so as to ensure or expand their share of the market; they may introduce new products which expand their potential markets; they may take over other businesses. All these things have

[9] Notably the American writer, H. A. Simon.

TABLE 16

INDICES OF GROWTH IN SALES BY SUBSIDIARIES, 1956–7 TO 1967–8

(1956 = 100)

Year ending 31 March	Dorman and Smith	DS Plugs	British Klockner	John Booth Foundries	Alorite
1957	115	107	106	128	164
1958	116	109	110	140	138
1959	116	103	110	130	214
1960	131	139	125	154	280
1961	149	137	152	159	190
1962	180	162	178	171	165
1963	190	194	183	164	228
1964	204	234	176	171	201
1965	249	238	193	228	243
1966	280	218	189	257	124
1967	290	212	162	276	89
1968	388	241	147	248	190

contributed in the Dorman Smith group.

The growth of sales of circuit breakers has provided a steady upward trend in sales. The company was in at the beginning of the English market, and has benefited from the growing but still incomplete acceptance of circuit breakers. It has maintained a continuous development programme, raising the capacity of miniature circuit breakers and developing moulded case circuit breakers, notably the *Loadline* series announced in 1962, following development work by Geoffrey Atherton, W. E. Minoprio and Harry Grantham who is now chief designer. The series has now been developed to deal with currents up to 2,500 amps, which is far beyond the capacity of any previous Dorman Smith equipment. Moreover, in the market for circuit breakers, a number of companies have entered since Dorman Smith, often with backing from American companies, but they have failed to challenge seriously. Ironically, one of the American companies involved was Federal Electric, with whom Geoffrey Atherton worked for a year. Thus in its key lines, the group has had technical and commercial strength.

The group has also introduced entirely new products. Some succeeded and others failed, but the net effect has certainly been positive.

A wide variety of new products has been considered by the member companies of the group. Several products, such as a washing up brush, an electric knife sharpener, electric razors and street lighting were seriously examined and potential overseas products tested with a view to production under licence, only to be rejected if there were serious problems of production in Britain or serious doubts about its suitability for the British market or the ultimate size of that market. A determined attempt was made to enter the market for domestic consumer durable goods at the beginning of the decade after Oldfield Engineering Company was renamed Dorman and Smith (Domestic). Many types of washing machine had been tested at the Preston works during 1958 and 1959, but the company could not secure the co-operation of the German company whose design it preferred, nor were its marketing consultants optimistic about prospects. The decision was taken not to proceed, which today seems an even more sensible choice in view of the subsequent growth of Italian manufacture. However, it was decided to proceed with a simpler clothes dryer, and production began in September 1959. Sales were promising at first, but slumped in 1960. Unfortunately, a similar dryer had been introduced by a Mr A. J. Flatley and severely undercut the Dorman Smith product. Dorman Smith cut the price of their dryer to just under £7, which did not cover overheads, and still failed to make enough headway in the market, and production was abandoned in May 1960. Later, one explanation of Flatley's low prices was revealed when he became bankrupt in 1962 and was imprisoned in the following year.

A more successful innovation arose out of a visit to the United States by Geoffrey Atherton. In March 1966 he wrote a memorandum to his fellow directors after driving through California:

> I noticed the widespread use of transistorised, battery-operated traffic flashers. These appear to have completely superseded the old paraffin lamps that surrounded holes in the road in this country. I obtained a sample made by a Los Angeles firm and telephoned Mr John Dietz of the R. E. Dietz Company of Syracuse, New York.
>
> I consider that this type of product might well find a substantial market in the United Kingdom. Road building and alterations will be a considerable industry in this country for many years ahead . . .

Dietz agreed to license Dorman and Smith to manufacture the flashers in Britain, and sold a batch of five hundred to the company to help evaluate the potential of the British market. However, there was a considerable element of risk, as Geoffrey Atherton committed the group to royalties on the basis of a six-figure turnover within two and a half to three years. His company had the manufacturing skill to produce the lamp, but the marketing skill had to be proven. The trade name TrafiLAMP was adopted because the American Visi-Flash had already been registered in Britain. The first order came from the Lancashire County Constabulary and exports began quickly to Australia. Their use spread rapidly as it became appreciated that the higher purchase cost of the lamp compared with the traditional paraffin lamp was far offset by the lower operating and maintenance costs. Geoffrey Atherton's risk paid off and developed a significant extra turnover for the group – enough to merit the establishment of a Hazard Lighting Division. Variations on the theme followed, with brighter daylight lamps and recently the TwinFLASH school crossing light, so that now these items are probably the most commonly seen products of the group.

Once the marketing of hazard lamps was established, other possibilities arose. In March 1968 another American product caught the eye of Stanley Thompson, manager of the Hazard Lighting Division, while he was visiting the United States; this was a reflective road stud – in effect a solid catseye – which was about half the cost of the compressible metal and rubber one and could be glued on to the road surface cheaply and quickly with resin instead of requiring a special recess. Moreover, its reflective qualities were superior to the catseye. Atherton discovered that there was only one rival in Britain, importing, rather than manufacturing, a similar product. In October 1968 a licence agreement was signed with the Reflex Corporation of Canada and production began within a month. Approval of the stud by the Ministry of Transport was slower than the company wished, but the wheels of officialdom are not geared to the urgency of building up turnover on items produced under licence; eventually, however, it was approved and as well as contributing to the group's profits, it benefits the motoring public since it is now economic to lay studs where catseyes were considered too expensive. In 1970 the Hazard Lighting Division merged with Dorman and Smith (Domestic) to form a new subsidiary, Dorman Smith Traffic Products Limited, which now exports a very significant percentage of its turnover.

A less successful venture recently was the Reinforced Plastics Division, set up in 1969 to make plastic housings for electrical controls and switchgear, and for a variety of outside customers. However, despite a flow production line, costs remained high and in February 1971, as Dorman Smith Reinforced Plastics Limited, it was sold to another plastics company. This released considerable factory space and BK Switchcontrols Limited (formerly British Klockner Switchgear) was transferred to Preston in April 1971 and the Chertsey factory was sold.

As well as sales, there have been acquisitions. In 1962 C. H. Parsons Limited and its subsidiary, Britmac Electrical Company, were bought together and Eric Atherton was appointed their managing director. The group was more interested in the goodwill of the business than its physical assets, for it made a wide range of electrical accessories such as switches, plugs, sockets and bell pushes, including a high quality range that carries the Royal Warrant. The Birmingham works and plant were sold and manufacture transferred to Preston. In 1968 production was transferred to the newly acquired Progress Mill on Shelley Road, Preston, a factory of 46,000 square feet. A number of products were considered for addition to the company range: burglar alarms and an inexpensive stereo amplifier were considered but recently nurse-call and bed-head units have been successfully introduced and added to the business of the subsidiary, which had been rather disappointing in the early days.

In the reorganisation of 1970, DS Plugs was split up, the plug interests being added to Britmac to form Dorman Smith Britmac Limited and the fuse manufacture at Salterbeck becoming Dorman Smith Fuses Limited.

Overseas performance

Overseas, the group repurchased an interest in the South African company in 1964. Desmond Tangney bought Dorman Electrical Sales Company (Proprietary) Limited in 1955 and built up a worthwhile business, though he was anxious to find a buyer for his partner's share of the business. As he was selling some circuit breakers imported from the United States, the directors of the British group felt that this was bad for the company image and bought an 18.9 per cent interest in Tangney's company to ensure a continuity of outlet for the group's products. A price was agreed, but after the bitter lesson of earlier experience, a minimum dividend was stipulated for the first two to three years. An option to buy more shares was only partly exercised and the group interest now amounts to 21.21 per cent of the issued capital. In most years, it has contributed satisfactory dividends to the group and even when it has not, it has provided a significant and profitable outlet for the British products.

The Indian interests in Hindustan Klockner and Morarji Dorman Smith have not been quite as successful. While they have contributed to group profits, the British board feels that their capital would probably be more profitably employed at home, especially as profitability has been increasing here in recent years. In 1971 therefore, the sale of the Indian investment was considered, and negotiations were well advanced by mid-1972.

Exports in recent years have been extremely successful, as the figures for the group show:

TABLE 17
SALES AND DIRECT EXPORTS, 1967–8 TO 1970–71

Year to 31 March	Sales (£)	Direct exports (£)	Percentage exported (per cent)
1968	2,322,621	256,288	11.0
1969	2,917,845	367,061	12.6
1970	3,714,496	471,840	12.7
1971	4,485,881	716,085	16.0

Moreover, the export figures take no account of indirect exports when Dorman Smith products are incorporated into the products of other companies' exports, as they commonly are.

Although the group exports to many countries (including Russia and the United States) in all five continents, the pattern of trade has changed in recent years. The major markets in the first half of the century were undoubtedly imperial, but the changing commercial climate was noted by the companies' executives in the sixties. In 1964 Thomas Atherton announced a sales drive to countries outside the British Commonwealth[10] which yielded a fivefold increase in exports to those countries within a year.[11] He had already expressed his confidence in the group's ability to compete in Europe in his chairman's statement of 1963, when he regretted the rejection of the British application to join the Common Market.

Two reports to the directors in March 1966 make a pointed contrast. Following a visit to France and Germany, Denis Kinnell reported:

> . . . the Continental markets would be important to the company . . . it was absolutely necessary to back their (agents) efforts in these Continental markets in order to find a replacement for the old

[10] *The Electrical Review*, 31 July 1964, p. 192.
[11] *The Electrical Review*, 27 August 1965, p. 325.

> Commonwealth markets which had little or no future due to the degree of protection being granted in those markets to local manufacturers.

Following a visit to Australia, New Zealand and the United States, Geoffrey Atherton wrote of Australasia:

> The degree of protection for locally manufactured goods is horrifying. I seriously doubt whether we should continue to regard Australia as a market worth much extra effort. I am also horrified by the degree of protection given to local New Zealand industry, and by the tendency to refuse British and buy Australian. For all practical and commercial purposes the Commonwealth no longer exists.

Since then there has been some debate in Australia about the long run desirability of very high and widespread protection of industry, but the lessons for Dorman Smith were clear. Despite the Common Market tariff, the biggest single export market in 1971 was France, and the company is confident on the threshold of Britain's entry into the Market.

Personnel

The nineteen sixties not only witnessed a fundamental change of emphasis in export markets, but also the end of one era for Dorman Smith and the beginning of another. On 1 July 1964, Thomas Atherton retired as managing director of Dorman and Smith. His son, Geoffrey, took over and as the events of this chapter show he has proved to be a worthy successor. It is, in fact, remarkable to find two generations so able and dedicated in a business as these two, and it is impossible to say which one history will consider to have made the greater contribution. Questioned, son would nominate father; and father, son. Thomas Atherton started with virtually nothing and successfully led the company for many more years than Geoffrey has done so far; yet the father is justifiably impressed by the remarkable success of the group in recent years under Geoffrey's leadership. But there is no doubt that without them, Dorman Smith Holdings would be a mere fraction of its present size and that the history of companies cannot be reduced to balance sheets and ratios, but must consider the personalities involved. The contrast between Baggs and the Athertons is a most striking illustration of this.

On 10 January 1969 Thomas Atherton resigned his chair on the board in favour of Geoffrey. He had served forty-four years with the company, entering as cashier and serving as chairman for thirty-two. He transformed it from an ailing elderly enterprise into a modern medium sized enterprise, employing 1,200 people, with a full quotation of its ordinary shares on the stock exchange, to which we shall return presently. Thomas Atherton still lives in Southport with his wife, Margaret, and he is able to enjoy his hobbies of reading and cricket, for as a keen supporter of the Red Rose he has served proudly on the committee of Lancashire Cricket Club since 1958 and been instrumental in gaining a new pavilion for the Southport ground. He retains an office at the Atherton Works and takes an active interest in the business that owes so much to him. His aims of 1937 could hardly have been more completely fulfilled: the company, quoted and grown out of all recognition, yet still in 1971 in the control of the family with his two sons holding forty-two per cent of the voting stock.

Other stalwarts of the company have also retired with the passage of time. In 1967 Robert Dale was succeeded as secretary by K. F. Blackshaw, who had been assistant secretary since 1962. Bruce Cooper retired from his executive duties and managing directorship of DS Plugs in August 1969 but still sits as vice-chairman of Dorman Smith Holdings. Gilbert Standen retired from BK Switchcontrols in 1969, though both he and Dale retained a seat each on one of the subsidiary boards. New blood entered the company, notably with the arrival of G. W. Barlow in 1969; he had been with the English Electric Company but left after Arnold Weinstock of GEC took the company over in 1968.[12] He was widely known and respected in the electrical industry

[12] The interesting story of this merger is told by Robert Jones and Oliver Marriott, *Anatomy of a Merger*, Jonathan Cape, London, 1970.

and his recruitment provided valuable outside experience and considerable ability at a time when the group needed it, though it was recognised that there was an inherent risk that a man of his capabilities was likely to be offered even more attractive posts outside. He was given a wide brief to explore new opportunities for the group. However, in 1970 he accepted an invitation to become chief executive of the new ball and roller bearing company, Ransome Hoffman Pollard. He continues as an outside director however, taking a positive interest in Dorman Smith and providing useful contacts in industry. For example, Hans Breitenbach, the director of manufacturing and engineering of Dorman Smith Switchgear, was recruited by him.

Generally, industrial relations within the group have been good. Geoffrey Atherton was especially impressed during his time in the United States by the extent to which an appreciation of the concept of economic efficiency pervaded the shop floor, and has tried, without reproducing it entirely, to generate the same spirit at Preston. Some of the credit for this belongs to Mr K. B. Glassby, who first appeared as one of Associated Industrial Consultants' employees at the Ordsal Works in 1954. The more senior staff at the Atherton Works were often loyal employees of the company, ex Ordsall, and were committed to the company, which in 1963 had increased the staff pension to fifty per cent of final salary. On the shop floor however, industrial relations were less than ideal. Glassby was invited on to the board of Dorman and Smith Limited in 1968 as director of manufacturing; he had not only had experience as an industrial consultant but had also been managing director of Rylands and Whitecross and director of Lancashire Steel Manufacturing Company and Lancashire Wire Company.

In the fifties, Glassby had set up work measurement techniques at Ordsall but these had gone out of date and much of the new work that had grown up over the years was not effectively measured. Resident work study staff found that the complexities of assessing work time on the very wide range of products[13] was

[13] The types of products have only been described very generally in this work but many of them were produced in a very wide range of capacities, so that the numbers of different items made were very high.

absorbing all their time. Glassby introduced an establishment scheme to try and simplify matters.[14] He was appointed to the board of Dorman Smith Holdings but left later in 1969 to set up his own consulting firm. Though his scheme continues at two of the three Preston factories, it has not been applied throughout the whole of the group; however, it has proved useful in reducing absenteeism and time losses generally.

So far as we can ascertain, there have only been two strikes in the history of the company. One was a very brief stoppage at the Salterbeck Works in its early days, the other in the Shelley Road works at Preston in February 1970 when fifty-six members of the AEF went on strike, alleging that a female worker had been put on measured work. It was settled later and agreed that non-union members should join the Municipal and General Workers' Union.

Shareholders

Although the preference shares had been quoted in 1948, the ordinary shares remained effectively in the hands of the Athertons, Coopers and a small number of selected employees. By 1962 Thomas Atherton was anxious to capitalise some of his investment and spread his assets so as to minimise liability to death duties. By offering some of the ordinary shares on the stock exchange, he could create a market for ordinary shares so that his death would not precipitate a forced sale of shares; and he could use the cash proceeds to buy agricultural land, which had estate duty advantages. Consequently, he and his family are now substantial owners of land around Southport.

In early 1962 a bonus issue was authorised, which gave thirteen ordinary shares and thirteen 'A' non-voting ordinary shares for every fourteen ordinary shares already held. Shortly afterwards there was an offer for sale of 353,000 units – each unit comprising one ordinary share which carried voting rights and one 'A' non-voting ordinary share. The units therefore had a nominal value of eight shillings but the offer price was 23s 6d per unit, and they

[14] It is described at length in the *Financial Times*, 10 September 1969, p. 13.

formed just over twenty-eight per cent of the total ordinary capital, so that there was no question of the existing management losing control. As there has been no foreseeable likelihood of any change of control, the value attributed to voting rights has been small, with little difference between the price of ordinary and 'A' shares. Since the ordinary shares were offered to the public there have been four bonus issues, in each case giving non-voting shares; the distributions were as follows: and an equivalent holding if one bought five shares in 1962 is shown in brackets:

1964 1 'A' share for every 5 ordinary held (giving 6 shares)
1968 1 'A' share for every 5 ordinary held (giving 7.2 shares)
1970 1 'A' share for every 5 ordinary held (giving 8.6 shares)
1971 1 'A' share for every 5 ordinary held (giving 10.32 shares)

Hence, if we bought 5 units of two shares each in 1962 at the offer price of 5 × 23s 6d, we should have paid £5 17s 6d for an investment worth approximately £30 at the beginning of September 1971 when the 'A' shares were quoted at £1.45. This is an increase of 457 per cent compared with an increase of 38.5 per cent in the *Financial Times* index of ordinary share prices over the same period. The fact that Dorman Smith is still controlled by one family, with Geoffrey and Eric Atherton as substantial ordinary shareholders, means that one of the main objections to the modern corporation – that the interests of shareholders and directors are not necessarily synonymous – cannot be applied in this case. The only exception to this harmony might arise if there were a takeover bid for the company, but the power of the family holding and the current good performance and high price earnings ratio make this an unlikely possibility.

Thus the shareholders have fared well, and short of any disaster, may continue to do well in the future. Certainly the implication of the stock market valuation is that investors expect continued growth in the future.

The future

Dorman and Smith is not a large firm by British standards[15] and there are many bigger ones in the electrical industry. It is commonly said that only the giant corporations can survive in an era of technological change and economies of scale. What then of the prospects of a firm like Dorman and Smith? First, it should be said that the company *has* existed in the shadow of giant electrical companies since the middle 1880s when Raworth sold out to Dorman and Smith and went to work for Brush, one of the giants which has since been swallowed up. Second, many economists consider that the case for 'bigness' as a paramount virtue is far from proven and that provided a firm can be efficient in its chosen line, it is not necessary to pursue size for its own sake. In its main activity of making switchgear (which accounted in 1970 for seventy per cent of turnover), the group has withstood strong efforts at competition and established itself as a market leader. Geoffrey Atherton observed in 1970:

> We have now got to the size where we don't have to change our policy to every shift in policy of other companies in the country. We have always regarded the company first and foremost as a commercial enterprise: the fact that we happen to make electrical equipment I didn't and still don't regard as being something that we're stuck with forever. For example, in our traffic products, I thought it would be a good idea to try a product for which we had the right degree of technical manufacturing skill, see if we could use it, and build up the selling skill. It's been an interesting exercise and I think it helped to prove to me that if we ever did find the electrical industry on the decline, we could hope to succeed in the manufacture and sale of other products.

The management has proved its competence in recent years, captured the lead in its field and has a very clear image of itself.

[15] In *The Times 1000* companies, which ranks manufacturing companies by their turnover, Dorman and Smith was number 930 in 1970–71.

PLATE 27. Dorman Smith TrafiSTUDS, providing better reflecting qualities at half the cost of catseyes, are now used on Britain's motorways

PLATE 28. The board of directors of Dorman Smith Holdings Limited and subsidiary companies, 1972. *Standing, left to right*: A. D. Gratton, S. M. Thompson, C. Bickerton, P. T. Beniston, K. F. Blackshaw, D. Kinnell, R. G. Dale, G. A. Standen, H. W. Breitenbach. *Seated, left to right*: G. W. Barlow, E. J. Atherton, T. G. F. Atherton, B. L. Cooper, M. L. Cooper

Its costs of production are low enough to enable it to clear the existing Common Market tariff wall, and can only give confidence in its prospects after British entry, when the wall disappears. Investment has been at a high level; in 1970–71 when it was generally sluggish in British industry, Dorman Smith increased their plant and machinery by twelve per cent.

Perhaps the last word on the issue of size should be left to the chairman, who wrote in 1970:[16]

> One of the most interesting aspects of the future will be the performance of the giants of the industry. Will they, in what to them are fringe activities, be able to compete effectively with the well organised, efficient medium sized company?

[16] *The Electrical Review*, 2 January 1970, p. 8.

APPENDIX 1

DORMAN AND SMITH: LIST OF INSTALLATIONS

STEAMSHIPS

Owner	*Lamps*	
	Glow	*Arc*
INMAN STEAMSHIP CO, LIVERPOOL		
City of Berlin	—	20
City of Rome	212	16
City of Paris	—	8
City of Chicago	240	—
City of Chester	26	4
Messrs GUION & CO, LIVERPOOL		
Alaska	400	—
Arizona	300	—
PACIFIC STEAM NAVIGATION CO, LIVERPOOL		
Potosi	40	—
ORIENT STEAM NAVIGATION CO, LONDON		
Chimborazo	40	—
Orient	178	9
Austral	178	9
Austral (Refitted)	250	4
Garonne	50	—
Cuzco	70	—
Lusitania	120	—
BRITISH INDIA STEAM NAVIGATION CO, GLASGOW		
India	75	—
COMPANIA TRANSATLANTICA, CADIZ		
Antonio Lopez	70	—
Cataluña	90	—
Ciudad de Santander	90	—
COMPAGNIE GENERALE TRANSATLANTIQUE, HAVRE		
Normandie	390	16
Messrs DONALD CURRIE & CO, LONDON		
Hawarden Castle	68	1
Norham Castle	68	1
Roslin Castle	68	1

CUNARD STEAMSHIP CO, LIVERPOOL		
Servia	180	—
Aurania	550	2
GREAT EASTERN RAILWAY CO, HARWICH		
Norwich	50	—
Ipswich	50	—
Messrs HENDERSON BROS, LIVERPOOL		
City of Rome (Additions)	213	—
Messrs FLINN, MAIN & MONTGOMERY, LIVERPOOL		
Vancouver	300	—
Messrs ISMAY, IMRIE & CO, LIVERPOOL		
Germanic	416	—
LONDON & NORTH-WESTERN RAILWAY CO, HOLYHEAD		
Banshee	100	—
CHILIAN STEAMSHIP CO		
Maipo	—	3
BRAZILIAN GOVERNMENT		
Vital de Oliveira	—	1
Messrs WEHNER, ASHTON & CO, MANCHESTER		
Alejandro	74	—
AUSTRO-HUNGARIAN LLOYD STEAM NAVIGATION CO, TRIESTE		
Euterpe	140	—
NEW ISLE OF MAN STEAM NAVIGATION CO, LIVERPOOL		
Lancashire Witch	50	—

WORKS, SHOPS, HOUSES ETC

Messrs HENRY TATE & SONS, Love Lane, Liverpool		
Sugar Refinery	600	5
Messrs F. STEINER & CO, Accrington		
Dyeing and Printing Works	350	16
Messrs ELKINGTON & CO, Birmingham		
Electro-plating Works	100	8
BRIDGEWATER TRUSTEES, Walkden		
Offices	100	—
J. KERR, Esq, Dunkenhalgh Hall		
Private House	92	—
Messrs GEORGE DEAKIN LIMITED, Winsford, Cheshire		
Bostock Salt Works	57	—
Messrs LEWIS & CO, Manchester		
Retail Shop	50	33
Messrs J. STEWART & CO, Blackwall, London		
Engine Works	36	—

Messrs PELLING, STANLEY & CO, Liverpool		
Offices	20	—
BLAENAVON STEEL & IRON CO		
Blaenavon Steel Works	—	20
Messrs J. WILKES & SON, Birmingham		
Copper Works	—	12
Mr ARTHUR H. LEE, German Mill, Bolton		
Cotton Mill	—	1
Mr GEORGE MOSS, 54 South Hill Road, Liverpool		
Contractor	—	2
Messrs T. HALL & CO, Birmingham		
Retail Shop	—	14
Arcade	—	3
T. RICHARDSON, Esq, MP, Yarm, Yorks		
Private House (Accumulator etc.)	—	—
J. HOPKINSON, Esq, Huddersfield		
Arcade	—	1
Messrs BINYONS, ROBINSON & Co, Manchester		
Retail Shop	—	1
Messrs LEWIS & CO, LIVERPOOL		
Retail Shop	—	2
SOUTHPORT WINTER GARDENS CO		
Conservatory and Pavilion	—	8
LONDON & NORTH-WESTERN RAILWAY CO		
Holyhead Harbour	—	5
Messrs LAIRD BROS, Birkenhead		
Graving Dock	—	9
J. BROOKS, Esq, Hepworth, Huddersfield		
Dyehouse	—	1
Messrs E. BRISTOWE & CO, Hull		
Warehouse	—	1
J. RILEY, Esq, Manchester		
Pomona Palace	—	1
Messrs L. DEMUTH & CO, Oldbury		
Chemical Works	—	1
Messrs D. ROLLO & SONS, Liverpool		
Foundry	—	12
Graving Dock	—	6
Messrs TANGYES LIMITED, Birmingham		
Steel Works	—	3
HENRY BODDINGTON, Esq, Pownall Hall, Wilmslow		
Private House (in Progress)	—	—

APPENDIX 2

CIRCULAR LETTER FROM DORMAN AND SMITH

Dorman & Smith, Electrical & General Engineers and Electricians
Manufacturers of Plant & Fittings
Successors to John S. Raworth
Telegraphic Address: 'Current', Manchester

24, Brazennose Street,
Manchester

We have much pleasure in laying before you some few data, bearing on the subject of the Illumination of Private Houses by means of the Electric Light.

We are strongly of opinion, that if the many advantages of the system were better known, it would be much more widely used than is at present the case. You will see from the enclosed letter to the Electrician, which is only one of many that can be quoted, that even an installation which was put down as long ago as 1881, when the system was quite in its infancy, has turned out eminently satisfactory. At the present time all the details of the system have been thoroughly worked out, and all the difficulties practically overcome. Installations on an extended scale cannot well be carried out until the Electric Lighting Act of 1882 is repealed or altered, but there is no cause why the residents in country and other detached houses, should not at once adopt what must be the illuminant of the future.

Of the many advantages offered by the system we would point out the following:—The light itself is as perfect as any artificial light can be. It is steady, soft, and of any desired brilliancy. It gives off no fumes, and very little heat; hence it may be used to any extent, without compunction, in drawing rooms, etc., without fear of damaging painting, flowers, or the most delicate fabrics, and, what is of much more importance, without fear of injuring the health of the occupants. It is very handy. This is especially the case when the dynamo is supplemented by an accumulator, in which case a light can be obtained in any part of the house, at any time of the day or night, by simply touching a switch, which can be placed in any convenient position—for instance, near the door in a dwelling-room or close to the bed in a bedroom. It gives almost absolute safety from fire, abolishing the risk of explosion, inseparable from the use of gas, and the danger of overturned lamps and candles, which has been the cause of so many destructive fires.

To many people who would otherwise at once adopt the Electric Light, the fact that some kind of engine must be put down to drive the dynamo, appears an insuperable objection. In practice there are few cases in which the difficulty is of great moment. Where water power is available it may be used with great advantage, as is shown by the subjoined letter. If gas is laid on, a gas-engine, in conjunction with dynamo and accumulator, makes an excellent installation. The engine can be placed in a cellar or outhouse, and allowed to run for a few hours during the day time without any special attention. The accumulator will then be available for the lamps at any time. For small installations where gas is not available, a low pressure engine—that is, an engine driven by steam at about the same pressure as the atmosphere—can be put in, with self-feeding boiler. All risk of boiler explosion is in this case entirely eliminated. Even if the installation is so large as to necessitate a high-pressure engine, the difficulties connected with working it are by no means great if it is properly designed for the work. Circumstances might permit, in many cases, of two or more householders in the same neighbourhood working their lights from one common plant. In the case of an installation without accumulator, and where of course it is necessary that the motor should continue running as long as any lights are required, a switch can be arranged in any part of the house, at any distance from the engine room, say in the principal bedroom, which in the case of gas-engine, turbine, or low-pressure engine, can be touched after the master of the house has retired to rest, and the motor, and with it the dynamo, brought to a stand still. The whole of the lights in the house will then be extinguished.

If you should wish to ascertain the probable cost of an installation on your premises, we shall have great pleasure in going into the matter for you, if you will let us know the probable number of lamps, the size of the premises, and probable position of dynamo.

We may say, for your guidance, that the ordinary 16 c.p. glow lamp gives out about the same amount of light as a five-foot gas burner.

The cost of running the lamps should be about a farthing per lamp per hour, in cases where the motor chosen does not require much attention; and should not exceed a halfpenny per lamp per hour, when an engine is used requiring an attendant.

Having been engaged in the development of incandescence lighting from its commencement, we can decide at once upon receipt of particulars what plant will be most suitable for any particular case.

We manufacture the necessary fittings and appliances in our own workshops, and are in a position to obtain engines, dynamos, etc., from all the best makers upon very favourable terms, not being tied down to any particular firm, we shall, in every case, be able to make the selection which our experience shows us to be the best. Having a staff of experienced workmen always in our employ, if you should entrust us with any work for you, we feel the utmost confidence in being able to carry it out, with despatch, to your entire satisfaction.

We remain,

Your obedient servants,

DORMAN & SMITH

APPENDIX 3

INTRODUCTION TO THE DORMAN AND SMITH TRADE CATALOGUE OF 1891

WE take advantage of the issue of another edition of our catalogue, to state that our position in the electrical world is that of first hand manufacturers. This being the case, we need make no apology, for leaving out of our price lists such items as engines and boilers, dynamo machinery, wire, and other necessaries, which we should have to buy from the firms to whom we advise users to go direct.

The goods scheduled in our lists are only our stock patterns; we make numerous others, and shall be glad to have enquiries from those who fail to find their requirements suited herein.

Although we do not pretend to do a merchanting business at home, or a shipping trade abroad in other than our own manufactures, we shall always be pleased, as heretofore, to give our foreign friends the benefit of our long experience in the selection of complete plants, for any system of electric lighting and transmission of power. To save useless enquiries, we may state that we do not traffic in bells, telephones, or telegraphs of any description, neither do we undertake any kind of installation work.

The electrical industry is now too large to allow of any one firm grappling successfully with all its sections; and the fact, of which we are justly proud, that with hardly an exception, all the leading firms of contractors and suppliers at home, on the continent, and in the colonies, take advantage of our experience and our facilities for manufacture, and place in our hands the work of making for them those goods in which we specialise, shows that there is a general concensus of opinion that the best work can only be turned out by special machinery, worked by trained hands under the supervision of experts.

To meet the ever increasing demand for our manufactures, we have acquired large works in Salford, and we expect by March next to have them in full swing. A view of the front factory is shown on the next page. To these works, which when finished, will be the most complete for our branch of the industry in the country, we intend moving our head offices on March 25th, 1892.

DECEMBER, 1891 DORMAN & SMITH

INDEX

Abbreviations: CMD – Charles Mark Dorman, HGB – Herbert George Baggs, JSR – John Smith Raworth, RAS – Reginald Arthur Smith, TA – Thomas Atherton, WTTB – W. T. Tallent Bateman.